UNDERSTANDING
THE GENOME

UNDERSTANDING THE GENOME

FROM THE EDITORS OF *SCIENTIFIC AMERICAN*

Compiled and with introductions by

George Olshevsky

Foreword by George M. Church, Ph.d.

a byron preiss book

WARNER BOOKS

An AOL Time Warner Company

Copyright © 2002 by Scientific American, Inc., and Byron Preiss Visual Publications, Inc. All rights reserved.

The essays in this book first appeared in the pages and on the web site of Scientific American, as follows: "The Human Genome Business Today" July 2000; "Where Science and Religion Meet: Francis S. Collins" February 1998; "An Express Route to the Gemone?: J. Craig Venter" August 1998; "Deciphering the Code of Life" December 1999; "The Unexpected Science to Come" December 1999; "Beyond the First Draft" September 2000; "Genome Scientists' To-Do List" September 2000; "Finding the Genome's Master Switches" December 2000; "Vital Data" March 1996; "Genetic Discrimination: Pink Slip Your Genes" January 2001; "The Bioinformatics Gold Rush" July 2000; "Hooking Up Biologists" November 2000; "Interview with Stuart Kauffman" June 2000; "Code of the Code" April 2001; "Beyond the Human Genome" July 2000; "The 'Other' Genomes" July 2000; "Reading the Book of Life" February 2001.

Warner Books, Inc., 1271 Avenue of the Americas, New York, NY 10020
Visit our Web site at www.twbookmark.com.

W An AOL Time Warner Company

Printed in the United States of America
First Printing: March 2002
10 9 8 7 6 5 4 3 2 1

ISBN: 0-446-67872-4
Library of Congress Control Number: 2001093742

Cover design by J. Vita
Book design by Gilda Hannah

Contents

, all new drugs must ultimat
ted in mammals—and that ofte
e. Mice are very close to h
ms of their genome: more tha
cent of the mouse proteins i
far show similarities to kno
teins. Ten laboratories acr
called the Mouse Genome Se
work, collectively received
m the National Institiutes o

Foreword

George M. Church

You opened this book at just the right time! Not just a time of discovery of hidden beauty about ourselves and our world, but a time when you must cast your vote on how you will perceive the message of genomes. As a society, will we try to hush the message, as with Darwin and Galileo? Or will we embrace it as we do the secrets of silicon, the key to electronics, or microbes, the key to public health? Will the genome message be so intuitive that we discuss it daily, like weather, psychology, and computers? Or will it be shrouded in the obscurity of complexity, like brain surgery, or for reasons of global security, like toxin manufacture?

In addition to the general perceptions, we have our individual concerns. Will our individual genomes be available for public view? Will viewing be optional, like our phone numbers, mandatory, like our faces, or forbidden, like our "private" anatomical features? The answers to these questions are being determined right now, so this is the ideal time to learn about genomes.

Ourcellsareminiaturelibrarieshousingfortysevenpreciouscryptically punctuateddrunonsentencesbeginnotechromosomesuptothreehundredmillionlettersofmainlyfourtypesacgtlongendnoteinterruptedfrequentlybeginnotebylongnotesoftenofgreatpracticalorhistoricalimportanceendnotalldedicatedtotheproliferationofselfassemblingnanomachines.

The text above dramatizes where scientists stand with the human genome. Now our job and joy is to decode and use these results wisely. But how do we begin to decipher the importance of billions of combinations of four letters (A,C,G,T)?

Right now, the barriers to progress are technical and societal. Knowledge of the genome is important both to the computational interpretation of those functional genomic data and to altering those functions. Technically, reading two bits (one base pair) of genetic data costs about two bits ($0.25), while reading two bits of data with a digital camera is a million times less expensive. Through advances in technology, the cost and speed of reading our genomes has dropped one thousand-fold in 15 years and will likely keep going, so this is a hurdle we can overcome.

Societally, our worries are more complex and include concerns about privacy and the price of progress. But how might our concerns about personal genomic codes change? Perhaps through increasingly focusing laws on "authorized use" of information rather than restricting access to it. Over time, individuals may even put their own medical information in publicly accessible web sites so that anyone can access it in times of emergency.

But this "transparency" leads to new questions about how this information will be used. Will human genes be altered to the point of creating designer babies? While changing human inheritance may become less controversial than it now seems, society might choose to avoid changing embryos and instead modify existing genes in adult bodies. Indeed, it is likely to be more effective to assess our full "genome function" well after birth than to extrapolate from or even complete genomic DNA information before birth. Engineering of adult cell genomes may one day become as routine as ways that we currently alter our bodies with cosmetics, drugs, vehicles, and education. In a sense, these adult changes are more readily "inherited" and spread than changes in embryos. That is, once a specific gene has been altered to the desired effect, everyone who wants the procedure can be treated in the same manner. However, genetic changes in embryos would take many generations before a wide portion of the population benefited from the alteration.

Scientists' advancing ability to assess individual mutations will change the way we are treated for disease, especially in the current one-size-fits-all drug market. Reprogramming of selected adult cells will likely replace the less precise practice of dosing our whole body with drugs. Over time, the idea that a small number of common mutations cause most common diseases (cancer, heart failure, psychiatric disorders) is likely to be replaced by the notion that the most harmful of our one million personal base pairs are recent (within the past 2000 generations) and rare (but common as a class).

Because the Human Genome Project will evolve to touch so many aspects of our lives, it may usher in a renaissance of interest in quantitative and statistical biology. Our enduring curiosity about ourselves may soon merge education, entertainment, and medicine such that books like the one that you hold open in your hands will be as ubiquitous as weather pages, gardening tips, and street maps.

Introduction

George Olshevsky

Although humanity did not know it at the time, the race to sequence the human genome began more than 135 years ago, with the publication of Gregor Johann Mendel's paper describing the results of his experiments on hybridizing pea plants. His experiments showed that the features of organisms were transmitted in discrete units of heredity. By 1926, biologist Thomas Hunt Morgan had named these units *genes*. Morgan's famous studies of the fruit fly, *Drosophila melanogaster*, located the genes on structures within the cell called *chromosomes*, which appear in the cell nucleus when the cell reproduces. These observations and experiments formed the foundation of *genetics*, the study of how the information needed to construct and maintain an organism is transmitted from generation to generation, how it is organized within the cell, and how it is expressed when the cell is alive.

Lacking a clear concept of information, nineteenth-century science did not grasp the idea that the features of an organism could be stored like a tape recording, copied, and transmitted to the succeeding generation. It wasn't until the 1940s that we learned that giant molecules of DNA, or deoxyribonucleic acid, served as the tapes, and not until 1953 that James Dewey Watson, Francis Harry Compton Crick, and Maurice Hugh Frederick Wilkins worked out the structure of DNA and how it functioned. The chromosomes were mainly DNA molecules all coiled up for the process of cell division, and the genes were comparatively short, well-defined sections along one or another of these DNA molecules.

Watson, Crick, and Wilkins, showed that each giant DNA molecule is a string of millions of base pairs, organized into hundreds or thousands of genes, that encode all the proteins that a cell can manufacture. At any moment in a given cell, thousands of ribosomes are churning out proteins, thousands of sections of DNA are separating and rejoining, and thousands of used proteins are being broken down into their component amino acids, which will quickly be reused to build other proteins. These processes go on ceaselessly in each of the trillions of cells in our bodies as long as we are alive.

The DNA molecules in our cells today originated more than 2.5 billion years ago, when cells with nuclei first appeared on earth. Their information has been passed down from generation to generation through the ages. They are very old molecules, and they are battle-scarred from their immense voyage through time. Their genes have been shuffled around, mutated, transposed, rearranged, duplicated, reduplicated, and transmuted, and the DNA we now possess bears only the vaguest resemblance to what it was like at the beginning. Nevertheless, by comparing our own DNA and the proteins encoded therein with those of other organisms which have undergone different transformations as those organisms evolved, we may discern our relationships to those organisms and, ultimately, discover our own evolution and our place in the scheme of nature.

And that is it in a nutshell: the Secret of Life.

Once this system was elucidated—and there is much work

still to do on the details—the Great Problem arose: to determine the complete sequence of base pairs in all the DNA molecules of a particular organism. This sequence is the organism's *genome*. Every individual organism, whether it is a bacterium, a plant, a worm, a salamander, or a human, has a unique genome; only clones and organisms that reproduce by cloning can have identical genomes. Bacteria and viruses have very simple genomes compared with those of multicellular organisms, so naturally their genomes were the first to be sequenced. Despite their simplicity, it took years of painstaking work to sequence those first genomes, but since then novel biochemical techniques have been developed that greatly reduce the sequencing time. By the late 1980s it became feasible to consider sequencing the entire human genome, which has a staggering 2.9 billion base pairs that organize themselves into 46 chromosomes.

In view of the enormous medical benefits and consequent money-making potential that will spring from knowing the human genome, a huge amount of politics inevitably accompanied the Human Genome Project from its inception. When the smoke cleared, two independent groups of laboratories, one commercial and one publicly funded, emerged as the primary sequencers. Each has its own method of doing the work, so they provide an independent cross-check of each other. In mid-February 2001, Celera Genomics, the commercial sequencer led by J. Craig Venter, and the Human Genome Project consortium, the publicly funded group led by Francis S. Collins, announced their "first drafts" of the human genome. At this writing, the sequence is more than 90% complete.

Now that a nearly complete human genome sequence is known, we must still determine what most of those genes (almost 40,000 of them) actually do. As we learn how to mine the information in the genome, we will become able to tailor medical treatments exactly to a person's genetic makeup, greatly diminishing the problem of unwanted drug side effects.

We will become able to cure, or at least to work around, devastating hereditary disorders. And we could in time become able to create nearly perfect "designer" babies, thereby ultimately finetuning our own evolution. The significance of the Human Genome Project to the human species is impossible to overstate.

The essays in this book of the *Scientific American* series describe various aspects of decoding the human genome. Now that the genetic code of a human being has been mapped, the first essay by Kathryn Brown details what researchers plan to do with it. Sir John Maddox, long-time former editor of the British science journal *Nature*, places the Human Genome Project into the overall context of science and attempts to describe what science might be like 50 years from now. Other essays describe the leaders of the two human genome sequencing groups, Celera and the Human Genome Project. One essay, coauthored by Francis S. Collins, discusses the overall significance of the Human Genome Project. Other essays deal with the business and pharmaceutical aspects of what we might do with the human genome once it has been completely sequenced. This includes applying the sequencing techniques used in the Human Genome Project to the genomes of other important organisms, such as laboratory animals and farm animals and plants.

The door into a "brave new world" of biology and medicine is now open—it will be fascinating to see what comes through.

Once we have the genome sequence, what do we do with it? The most immediate benefits from the Human Genome Project will be medical. Large pharmaceutical companies are aligning with savvy genomic researchers with hopes of creating genetically tailored medications and developing cures for currently untreatable illnesses. However, this rush to profits has raised many technical, legal, and social questions about who owns the variations of the the DNA chemicals that make up our genes.

The Human Genome Business Today

Kathryn Brown

Right now you can read the entire genetic code of a human being over the Internet. It's not exactly light reading—start to finish, it's nothing but the letters A, T, C and G, repeated over and over in varying order, long enough to fill more than 200 telephone books. For biologists, though, this code is a runaway best-seller. The letters stand for the DNA chemicals that make up all your genes, influencing the way you walk, talk, think and sleep. "We're talking about reading your own instruction book," marvels Francis S. Collins, director of the National Human Genome Research Institute in Bethesda, Md. "What could be more compelling than that?" The race to map our genes has cast a spotlight on the human genetic code—and what, exactly, researchers now plan to do with it.

"For a long time, there was a big misconception that when the DNA sequencing was done, we'd have total enlightenment about who we are, why we get sick and why we get old," remarks geneticist Richard K. Wilson of Washington University, one partner in the public consortium. "Well, total enlight-enment is decades away."

But scientists can now imagine what that day looks like. Drug companies, for instance, are collecting the genetic know-how to make medicines tailored to specific genes—an effort called pharmacogenomics. In the years to come, your pharma-cist may hand you one version of a blood pressure drug, based on your unique genetic profile, while the guy in line behind you gets a different version of the same medicine. Other com-panies are already cranking out blood tests that reveal tell tale disease-gene mutations—and forecast your chances of coming down with conditions such as Huntington's disease. And some scientists still hold out hope for gene therapy: directly adding healthy genes to a patient's body. "Knowing the genome will change the way drug trials are done and kick off a whole new era of individualized medicine," predicts J. Craig Venter, presi-dent of Celera.

Even with the human code in hand, however, the genomics industry faces challenges. Some are technical: it's one thing to know a gene's chemical structure, for instance, but quite another to understand its actual function. Other challenges are legal: How much must you know about a gene in order to patent it? And finally, many dilemmas are social: Do you really want to be diagnosed with a disease that can't be treated—and won't affect you for another 20 years? As scientists begin unraveling the genome, the endeavor may come to seem increasingly, well, human.

The "Race"

In the spring of 2000, all eyes were on the first finish line in the genome: a rough-draft sequence of the 40,000 or so genes inside us all. The HGP's approach has been described as painstaking and precise. Beginning with blood and sperm cells, the team separated out the 23 pairs of chromosomes that hold human genes. Scientists then clipped bits of DNA from every chromosome, identified the sequence of DNA bases in each bit, and, finally, matched each snippet up to the DNA on either side of it in the chromosome. And on they went, gradually crafting the sequences for individual gene segments, complete genes, whole chromosomes and, eventually, the entire genome. Wilson compares this approach to taking out one page of an encyclopedia at a time, ripping it up and putting it together again.

In contrast, Celera took a shorter route: shredding the encyclopedia all at once. Celera's so-called shotgun sequencing strategy tears all the genes into fragments simultaneously and then relies on computers to build the fragments into a whole genome. "The emphasis is on computational power, using algorithms to sequence the data," says J. Paul Gilman, Celera's director of policy planning. "The advantage is efficiency and speed."

The HGP and Celera teams disagree over what makes a "finished genome." In the spring of 2000, Celera announced that it had finished sequencing the rough-draft genome of one anonymous person and that it would sort the data into a map in just six weeks. But the public team immediately cried foul, as Collins noted that Celera fell far short of its original genome-sequencing goals. In 1998, when the company began, Celera scientists planned to sequence the full genomes of several people, checking its "consensus" genome ten times over. In its April, 2000 announcement, however, Celera declared that its rough genome sequencing was complete with just one person's genome, sequenced only three times.

Although many news accounts have characterized the HGP and Celera as competing in a race, the company has had a decided advantage. Because the HGP is a public project, the team routinely dumps all its genome data into GenBank, a public database available through the Internet (at www.ncbi.nlm.nih.gov/). Like everyone else, Celera has used that data—in its case, to help check and fill the gaps in the company's rough-draft genome. Essentially, Celera used the public genome data to stay one step ahead in the sequencing effort. "It does stick in one's craw a bit," Wilson remarks. But Gilman asserts that Celera's revised plan simply makes good business sense. "The point is not just to sit around and sequence for the rest of our lives," Gilman adds. "So, yes, we'll use our [threefold] coverage to order the public data, and that will give us what we believe to be a very accurate picture of the human genome." In early May the HGP announced it had completed its own working draft as well as a finished sequence for chromosome 21, which is involved in Down's syndrome and many other diseases.

Until now, the genome generators have focused on the similarities among us all. Scientists think that 99.9 percent of your genes perfectly match those of the person sitting beside you. But the remaining 0.1 percent of your genes vary—and it is these variations that most interest drug companies. Even a simple single-nucleotide polymorphism (SNP)—a T, say, in one of your gene sequences, where your neighbor has a C—can spell trouble. Because of these tiny genetic variations, Venter claims, many drugs work only on 30 to 50 percent of the human population. In extreme cases, a drug that saves one person may poison another. Venter points to the type II diabetes drug ezulin, which has been linked to more than 60 deaths from liver toxicity worldwide. "In the future, a simple genetic test may determine whether you're likely to be treated effectively by a given drug or whether you face the risk of being killed [that same drug," Venter predicts. While fleshing out its rough genome, Celera has also been comparing some of the genes with those from other individuals, building up a database of SNPs (pronounced "snips").

Other companies, too, hope to cash in on pharmacogenomics. Drug giants are partnering with smaller genomics-savvy companies to fulfill their gene dreams: Pfizer in New York City has paired with Incyte Genomics in Palo Alto, Calif.;

SmithKline Beecham in Philadelphia has ties to Human Genome Sciences in Rockville; and Eli Lilly in Indianapolis has links to Millennium Pharmaceuticals in Cambridge, Mass. At this point, personalized medicine is still on the lab bench, but some business analysts say it could become an $800-million marker by 2005. As Venter puts it: "This is where we're headed."

But the road is sure to be bumpy. One sticking point is the use of patents. No one blinks when Volvo patents a car design or Microsoft patents a software program, according to John J. Doll, director of the U.S. Patent and Trademark Office's biotechnology division. But many people are offended that biotechnology companies are claiming rights to human DNA—the very stuff that makes us unique. Still, without such patents, a company like Myriad Genetics in Salt Lake City couldn't afford the time and money required to craft tests for mutations in the genes *BRCA1* and *BRCA2*, which have been linked to breast and ovarian cancer. "You simply must have gene patents," Doll states.

Most scientists agree, although some contend that companies are abusing the public genome data that have been so exactingly sequenced—much of them with federal dollars. Dutifully reporting their findings in GenBank, HGP scientists have offered the world an unparalleled glimpse at what makes a human. And Celera's scientists aren't the only ones peering in—in April 2000, GenBank logged roughly 35,000 visitors a day. Some work at companies like Incyte, which mines the public data to help build its own burgeoning catalogue of genes—and patents the potential uses of those genes. Incyte has already won at least 500 patents on full-length genes— more than any other genomics company—and has applied for roughly another 7,000 more. Some researchers complain that such companies are patenting genes they barely understand and, by doing so, restricting future research on those genes. "If data are locked up in a private database and only a privileged few can access it by subscription, that will slow discovery in many diseases," warns Washington University's Wilson.

Incyte president Randal W. Scott, however, sees things differently: "The real purpose of the Human Genome Project (HGP) is to speed up research discoveries, and our work is a natural culmination of that. Frankly, we're just progressing at a scale that's beyond what most people dreamed of." In March 2000, Incyte launched an e-commerce genomics program—like an amazon.com for genes—that allows researchers to order sequence data or physical copies of more than 100,000 genes on-line. Subscribers to the company's genomics database include drug giants such as Pfizer, Bayer and Eli Lilly. Human Genome Sciences has won more than 100 gene patents—and filed applications for roughly another 7,000—while building its own whopping collection of genes to be tapped by its pharmaceutical partners, which include SmithKline Beecham and Schering-Plough.

The federal government has added confusion to the patent debate. In March 2000, President Bill Clinton and British prime minister Tony Blair released an ambiguous statement lauding open access to raw gene data—a comment some news analysts interpreted as a hit to Celera and other genomics companies that have guarded their genome sequences carefully. Celera and the HGP consortium have sparred over the release of data, chucking early talks of collaboration when the company refused to release its gene sequences immediately and fully into the public domain. The afternoon Clinton and Blair issued their announcement, biotech stocks slid, with some dropping 20 percent by day's end. A handful of genomics companies scrambled to set up press conferences or issue statements that they, indeed, did make available their raw genome data for free. In the following weeks, Clinton administration officials clarified that they still favor patents on "new gene-based health care products."

The sticky part for most patent seekers will be proving the utility of their DNA sequences. At the moment, many patent applications rely on computerized prediction techniques that are often referred to as "in silico biology." Armed with a full or partial gene sequence, scientists enter the data into a computer program that predicts the amino acid sequence of the resulting protein. By comparing this hypothetical protein with known proteins, the researchers take a guess at what the underlying gene sequence does and how it might be useful in developing a drug, say, or a diagnostic test. That may seem like a wild stab at biology, but it's often enough to win a gene patent. "We accept that as showing substantial utility," Doll says. Even recent revisions to federal gene-patent standards—which have generally raised the bar a bit on claims of usefulness—ask only that researchers take a reasonable guess at what their newfound gene might do.

Testing, Testing

Patents in July 2000 have already led to more than 740 genetic tests that are on the market or being developed, according to the National Institutes of Health. These tests, however, show how far genetics has to go. Several years after the debut of tests for *BRCA1* and *BRCA2*, for instance, scientists are still trying to determine exactly to what degree those genes contribute to a woman's cancer risk. And even the most informative genetic tests leave plenty of questions, suggests Wendy R. Uhlmann, president of the National Society of Genetic Counselors. "In the case of Huntington's, we've got a terrific test," Uhlmann avers. "We know precisely how the gene changes. But we can't tell you the age when your symptoms will start, the severity of your disease, or how it will progress."

Social issues can get in the way, too. After Kelly Westfall's mother tested positive for the Huntington's gene, Westfall, age 30, immediately knew she would take the test as well. "I had made up my mind that if I had Huntington's, I didn't want to have kids," declares Westfall, who lives in Ann Arbor, Mich.

But one fear made her hesitate: genetic discrimination. Westfall felt confident enough to approach her boss, who reassured her that her job was safe. Still, she worried about her insurance. Finally, rather than inform her insurer about the test, Westfall paid for it—some $450, including counseling—out of pocket. (To her relief, she tested negative.)

The HGP's Collins is among those calling for legislation to protect people like Westfall. A patchwork of federal and state laws are already in place to ban genetic discrimination by insurers or employers, but privacy advocates are lobbying Congress to pass a more comprehensive law. In February 2000, President Clinton signed an executive order prohibiting all federal employers from hiring, promoting or firing employees on the basis of genetic information. It remains to be seen whether private companies will follow suit.

In the meantime, Celera is now ready to hawk its human genome, complete with crib notes on all the genes, to online subscribers worldwide. "It's not owning the data—it's what you do with it," Venter remarks. He envisions a Celera database akin to Bloomberg's financial database or Lexis-Nexis's news archives, only for the genetics set. Which 300 genes are associated with hypertension? What, exactly, does each gene do? These are the kinds of queries Celera's subscribers might pose—for a price. As of this writing, Celera planned to offer a free peek at the raw genome data online, but tapping into the company's online toolkit and full gene notes will cost corporate subscribers an estimated $5 million to $15 million a year, according to Gilman. Academic labs will pay a discounted rate: $2,000 to $15,000 a year.

Internet surfers can now visit GenBank for free. With all this information available, will scientists really pay Celera? Venter thinks so. "We just have to have better tools," he says. For genomics, that is becoming a familiar refrain.

Thousands of people have contributed to the Human Genome Project (HGP). Their efforts must be paid for, collated, reviewed, corrected, and combined with the contributions of others, if the project is not to degenerate into chaos. As with any major undertaking, this requires a well-conceived organization and a management hierarchy. In the case of the HGP, there happen to be two such: the National Human Genome Research Institute (NHGRI) and The Institute for Genomic Research (TIGR), also known as Celera Genomics. This is a good thing, because the two organizations provide a cross-check on each other and thereby greatly diminish the number of sequencing errors. It is by no means a duplication of effort. Inevitably, however, it also engenders a competitive atmosphere, particularly for scarce resources. This may or may not be such a good thing, although so far it has kept things lively and moving right along. Not surprisingly, both projects are headquartered in the Washington, DC area, which is where the funding is located.

The heads of these two organizations are Francis S. Collins and J. Craig Venter. Both are staggeringly brilliant but quite different in their approaches to sequencing the human genome. The following two essays profile these two men and tell us something about what inspires them.

Where Science and Religion Meet: Francis S. Collins

Tim Beardsley

T he combination of world-class scientific researcher, savvy political activist, federal program chief, and serious Christian, is not often found in one person. Yet that constellation of traits is vigorously expressed in Francis S. Collins.

Collins leads the U.S. Human Genome Project, an ambitious effort to analyze the human genetic inheritance in its ultimate molecular detail. A physician by training, he became a scientific superstar in 1989, when he was a researcher at the University of Michigan. There, together with various collaborators, he employed a new technique called positional cloning to find the human gene that, if mutated, can give rise to cystic fibrosis. That discovery quickly made possible the development of tests for prenatal diagnosis of the disease.

Collins has since co-led successful efforts to identify several other genes implicated in serious illness. His tally of discoveries thus far includes genes that play a role in neurofibromatosis and Huntington's disease, as well as the rarer ataxia telangiectasia and multiple endocrine neoplasia type 1. In 1993, after turning down the invitation six months earlier, Collins left Michigan to become director of what is now the National Human Genome Research Institute.

In his office on the campus of the National Institutes of Health in Bethesda, Md., the 47-year-old Collins sits at the center of a vortex of medical hopes and fears that is probably unrivaled. He is widely seen as a strong leader for the genome program, which he reports is on target for sequencing the entire three billion bases of human DNA by 2005. And his influence extends well beyond research. Collins's energetic support for laws to prevent people from losing health insurance because of genetic discoveries is perhaps the best explanation for the limitations on gene-based insurance discrimination in the 1996 Kennedy-Kassebaum bill.

Recently Collins has thrown his political weight behind a new "potentially expensive but very important goal" that he hopes will supplement the genome project's sequencing effort. Collins wants to assemble a public-domain catalogue of subtle human genetic variations known as single-nucleotide polymorphisms, written "SNPs" and pronounced "snips." The effort would constitute "a very significant change in the vision of what the genome project might be," Collins says. SNPs are detected by comparing DNA sequences derived from different people.

Unlike positional cloning, analysis of SNPs can readily track down genes that, though collectively influential, individually play only a small role in causing disease. Diabetes, hypertension and some mental illnesses are among the conditions caused by multiple genes. New DNA "chips," small glass plates incorporating microscopic arrays of nucleic acid sequences, can be used to detect mutations in groups of genes simultaneously. By employing this chip technology, researchers should be able to use SNPs for rapid diagnoses.

Collins now spends a quarter of his time building support at NIH for a SNP repository. He bolsters his case by predicting that, absent a public effort on SNPs, private companies will probably survey these molecular flags and patent them. There may be only 200,000 of the most valuable SNPs, so patents could easily deny researchers the use of them except through "a complicated meshwork of license agreements."

Collins the federal official often retains the open-collar, casual style that is de rigueur among scientists, and his preferred mode of transportation (motorcycle) has earned him some notoriety. He is, however, more unassuming than officials or scientists are wont to be. He feels "incredibly fortunate" to be standing at the helm of a project "which he thinks is going to change everything over the years." Such feelings inspire Collins to musical expression. At the annual North American Cystic Fibrosis Conference, he performed his song "Dare to Dream," accompanying himself on guitar. Yet Collins's easygoing demeanor belies intensity not far below the surface: he estimates that 100-hour workweeks are his norm.

He grew up on a farm in Virginia and graduated with a degree in chemistry from the University of Virginia with highest honors. He followed up with a Ph.D. in physical chemistry at Yale University, then went to the University of North Carolina to study medicine. He was soon active in genetics. As a researcher at Michigan, he was doing "exactly what I wanted to do," which is why he turned down the job of leading the genome program the first time he was offered it. He now admits, however, he is "having a very good time."

Large-scale human DNA sequencing was not initiated until 1996, after preliminary mapping had been accomplished. The only cloud on the horizon that Collins foresees is reducing the cost enough to fit the entire project into the budget, $3 billion over 15 years.

Sequencing now costs 50 cents per base pair. Collins needs to get that figure down to 20 cents. If he could reach 10 cents, the gene sequencers could tackle the mouse as well, something Collins wants to do because comparisons would shed light on how the genome is organized. Cutting against that, however, is the need to ensure reproducibility. Collins has enacted cross-laboratory checks to ensure that sequence accuracy stays over 99.99 percent.

Collins notes with satisfaction that today there are people alive who would have died without genetic tests that alerted physicians to problems. Patients with certain types of hereditary colon cancer, which can be treated by surgery, are the most obvious examples. Testing for genes predisposed to multiple endocrine neoplasia type 1 and, possibly, breast and ovarian cancer may in time save lives, Collins judges.

Congress funded the Genome project hoping it would lead to cures. But for most of the diseases to which Collins has made important contributions, the only intervention at present is abortion of an affected fetus. Although normally fluent, Collins is halting on this subject, saying he is personally "intensely uncomfortable with abortion as a solution to anything." He does not advocate changing the law and says he is "very careful" to ensure that his personal feelings do not affect his political stance.

He volunteers that his views stem from his belief in "a personal God." Humans have an innate sense of right and wrong that "doesn't arise particularly well" from evolutionary theory, he argues. And he admits his own "inability, scientifically, to be able to perceive a precise moment at which life begins other than the moment of conception." Together, these ideas lead to his having "some concerns" about whether genetic testing and abortion will be used to prevent conditions that are less than disastrous, such as a predisposition to obesity.

The recent movie *Gattaca* thrust before the public eye the prospect that genetic research will in the near future allow the engineering of specific desirable traits into babies. Collins thinks it is "premature to start wringing our hands" about the prospect of genetic enhancement. But, he states, "I personally think that it is a path we should not go down, not now and maybe not for a very long time, if ever."

Researchers and academics familiar with Collins's work agree that he has separated his private religious views from his professional life. Paul Root Wolpe, a sociologist at the University of Pennsylvania, states that "[Collins's] history has shown no influence of religious beliefs on his work other than a generalized sensitivity to ethics issues in genetics." Leon E. Rosenberg of Bristol-Myers Squibb, a former mentor, says that "the fact that he wears his Christianity on his sleeve is the best safeguard against any potential conflict."

Despite the general approbation, Collins is not entirely without critics. John C. Fletcher, former director of the Center for Biomedical Ethics of the University of Virginia and an Episcopalian minister before he left the church, faults Collins for not pushing to remove the current ban on using federal funds for human embryo research. Research on early embryos could lead to better treatments for pediatric cancers, Fletcher argues.

In 1996 Collins endured what he calls "the most painful experience of my professional career." A "very impressive" graduate student of his falsified experimental results relating to leukemia that had been published in five papers with Collins and others as co-authors. After Collins confronted him with a dossier of evidence, the student made a full confession. But Collins thinks his feelings of astonishment and betrayal "will never fade."

The fraud was detected by an eagle-eyed reviewer, who noticed that some photographs of electrophoresis gels that appeared in a manuscript were copied. As a result, Collins says that when someone displays a film at a meeting, "instinctively now I am surveying it to see if there is a hint that something has been manipulated." Collins remarks that since the fraud became public, a "daunting" number of scientists have contacted him to describe similar experiences of their own.

Collins still runs his own laboratory, and he continues to press a "very sharp" policy agenda. These involvements keep him busy, but he will soon spend a month with his daughter Margaret, a physician, in a missionary hospital in Nigeria. During his last visit, almost ten years ago, he saved a man's life in a dramatic do-or-die surgery conducted with only the most basic instruments. These expeditions, to Collins, are an expression of his faith. But they are something else as well, he adds: "It seemed like it would be a wonderful thing to do with your kid."

e, all new drugs must ultim
sted in mammals—and that of
ce. Mice are very close to
rms of their genome: more t
rcent of the mouse proteins
far s imi to k
oteins. Ten laboratories a
called the Mouse Genome S
twork, collectively receive
om the National Institiutes

An Express Route to the Genome?: J. Craig Venter

Tim Beardsley

J. Craig Venter, the voluble director of the Institute for Genomic Research (TIGR) in Rockville, Md., is much in demand these days. A tireless self-promoter, Venter set off shock waves in the world of human genetics in May 1998 by announcing, via the front page of the *New York Times*, a privately funded $300-million, three-year initiative to determine the sequence of almost all the three billion chemical units that make up human DNA, otherwise known as the genome. The claim prompted incredulous responses from mainstream scientists engaged in the international Human Genome Project, which was started in 1990 and aims to learn the complete sequence by 2005. This publicly funded effort would cost about ten times as much as Venter's scheme. But Venter's credentials mean that genome scientists have to take his plan seriously.

In 1995 Venter surprised geneticists by publishing the first complete DNA sequence of a free-living organism, the bacterium *Haemophilus influenzae*, which can cause meningitis and deafness. This achievement made use of a then novel technique known as whole-genome shotgun cloning and "changed all the concepts" in the field, Venter declares: "You could see the power of having 100 percent of every gene. It's going to be the future of biology and medicine and our species." He followed up over the next two and a half years with complete or partial DNA sequences of several more microbes, including agents that cause Lyme disease, stomach ulcers and malaria.

The new, private human genome initiative will be conducted by a company, Celera Genomics, that will be owned by TIGR, a Perkin-Elmer Corporation (the leading manufacturer of DNA sequencers) and Venter himself, who will be its president. In March 2000 he knocked off the genome of the fruit fly Drosophila melanogaster, an organism used widely for research in genetics.

Venter has a history of lurching into controversy. As an employee of the National Institutes of Health in the early 1990s, he became embroiled in a dispute over an ultimately unsuccessful attempt by the agency to patent hundreds of partial human gene sequences. Venter had uncovered the partial sequences, which he called expressed sequence tags (ESTs), with a technique he developed in his NIH laboratory for identifying active genes in hard-to-interpret DNA. "The realization I had was that each of our cells can do that better than the best supercomputers can," Venter states.

Many prominent scientists, including the head of the NIH's human genome program at the time, James D. Watson, opposed the attempt to patent ESTs, saying it could imperil cooperation among researchers. (Venter says the NIH talked him into seeking the patents only with difficulty.) And at a congressional hearing, Watson memorably described Venter's automated gene-hunting technique as something that could be "run by monkeys." An NIH colleague of Venter's responded later by publicly donning a monkey suit.

Venter left the NIH in 1992 feeling that he was being

treated "like a pariah." And he does not conceal his irritation that his peers were slow to recognize the merits of his proposal to sequence *H. influenzae*. After failing to secure NIH funding for the project, Venter says he turned down several tempting invitations to head biotechnology companies before finally accepting a $70-million grant from HealthCare Investment Corporation to establish TIGR, where he continued his sequencing work. Today, when not dreaming up audacious research projects, Venter is able to relax by sailing his oceangoing yacht, the *Sorcerer*.

His assault on the human genome employs the whole-genome shotgun cloning technique he used on *H. influenzae* and other microbes. The scheme almost seemed designed to make the Human Genome Project look slow by comparison. To date, that effort has devoted most of its resources to "mapping" the genome—defining molecular landmarks that will allow sequence data to be assembled correctly. But whole-genome shotgun cloning ignores mapping. Instead, it breaks up the genome into millions of overlapping random fragments, then determines part of the sequence of chemical units within each fragment. Finally, the technique employs powerful computers to piece together the resulting morass of data to recreate the sequence of the genome.

Predictably, Venter's move prompted some members of Congress to question why government funding of a genome program was needed if the job could be done with private money. Yet if the goal of the Human Genome Project is to produce a complete and reliable sequence of all human DNA, says Francis S. Collins, director of the U.S. part of the project, Venter's techniques alone cannot meet it. Researchers insist that his "cream-skimming" approach furnishes information containing thousands of gaps and errors, even though it has short-term value. Venter accepts that there are some gaps but expects accuracy to meet the 99.99 percent target of the existing genome program.

Shortly after Venter's proposed scheme hit the headlines, publicly funded researchers started discussing a plan to speed up their own sequencing timetable in order to provide a "rough draft" of the human genome sooner than originally planned. Collins says this proposal, which would require additional funding, would have surfaced even without the new competition. Other scientists think Venter's plan has spurred the public researchers forward.

Venter has always had an iconoclastic bent. He barely graduated from high school and in the 1960s was happily surfing in southern California until he was drafted. Early hopes of training for the Olympics on the Navy swim team were dashed when President Lyndon B. Johnson escalated the war in Vietnam. But Venter says he scored top marks out of 35,000 of his navy peers in an intelligence test, which enabled him to pursue an interest in medicine. He patched up casualties for five days and nights without a break in the main receiving hospital at Da Nang during the Tet offensive, and he also worked with children in a Vietnamese orphanage.

Working near so much needless death, Venter says, prompted him to pursue Third World medicine. Then, while taking premed classes at the University of California at San Diego, he was bitten by the research bug and took up biochemistry. He met his wife, Claire M. Fraser, now a TIGR board member, during a stint at the Roswell Park Cancer Institute in Buffalo, N.Y., and took his research group to the NIH in 1984.

His painstaking attempts to isolate and sequence genes for proteins in the brain known as receptors started to move more quickly after he volunteered his cramped laboratory as the first test site for an automated DNA sequencer made by Applied Biosystems International, now a division within Perkin-Elmer. Until then, he had sequenced just one receptor gene in more than a decade of work, so he felt he had to be "far more clever" than scientists with bigger laboratories. Venter was soon employing automated sequencers to find more genes; he then turned to testing protocols for the Human Genome Project, which was in the discussion phase.

After leaving the government and moving to TIGR, Venter entered a controversial partnership with Human Genome Sciences, a biotechnology company in Rockville established to exploit ESTs for finding genes. The relationship never easy, floundered in 1997. According to Venter, William A. Haseltine, the company's chief executive, became increasingly reluctant to let him publish data promptly. Haseltine replies that he often waived delays he could have required.

The divorce from Human Genome Sciences cost TIGR $38 million in guaranteed funding. The day after the split was announced, however, TIGR started to rehabilitate itself with a suspicious scientific community by posting on its World Wide Web site data on thousands of bacterial gene sequences.

The sequencing building for Venter's human genome company, adjacent to TIGR, may become a technological mecca. It will produce more DNA data than the rest of the world's output combined, employing 230 of Perkin-Elmer Applied Biosystems 3,700 machines. These sophisticated robots, which sell for $300,000 apiece, require much less human intervention than state-of-the-art devices. Venter says the new venture will release all the human genome sequence data it obtains, at three-month intervals. It makes a profit by selling access to a database that will interpret the raw sequence data as well as crucial information on variations between individuals that should allow physicians to tailor treatments to patients' individual genetic makeups. Most of the federal sequencing centers do not look at the data they produce, Venter thinks, but just put it out as if they were "making candy bars or auto parts."

Celera Genomic will also patent several hundred of the interesting genes it expects to find embedded within the human genome sequence. Venter defends patents on genes, saying they pose no threat to scientific progress. Rather, he notes, they guarantee that data are available to other researchers, because patents are public documents. His new venture will not patent the human genome sequence itself, Venter states.

"We will make the human genome unpatentable" by placing it in the public domain, he proclaims. All eyes will be on Venter to see how closely he can approach that goal.

It just doesn't get any better than this: an essay on the Human Genome Project coauthored by one of the people in charge! It is like having Albert Einstein explain the Theory of Relativity, or Winston Churchill write about the role of England in World War II. Here Francis S. Collins and Karin G. Jegalian explain how sequencing the human genome will aid us in understanding life processes and will help to answer some of our most fundamental questions about the inner workings of the cell.

As the genome is sequenced, and for many years after, we will have to annotate it. That is, for each gene in the genome, we must determine its function or functions and explain how it interacts with the other genes. For a while, we will be overwhelmed with the quantity of data to be sifted through, but in time and with the advent of new data-processing and biochemical-structuring software, this will be done. As Collins and Jegalian point out, the annotated human genome will have profound effects on our society as well as on our science.

Deciphering the Code of Life

Francis S. Collins and Karin G. Jegalian

When historians look back at this turning of the millennium, they will note that the major scientific breakthrough of the era was the characterization in ultimate detail of the genetic instructions that shape a human being. The Human Genome Project—which aims to map every gene and spell out letter by letter the literal thread of life, DNA—will affect just about every branch of biology. The complete DNA sequencing of more and more organisms, including humans, will answer many important questions, such as how organisms evolved, whether synthetic life will ever be possible, and how to treat a wide range of medical disorders.

The Human Genome Project is generating an amount of data unprecedented in biology. A simple list of the units of DNA, called bases, that make up the human genome would fill 200 telephone books—even without annotations describing what those DNA sequences do. A working draft of 90 percent of the total human DNA sequence was in hand by the spring of 2000, and the full sequence is expected in 2003. But that will be merely a skeleton that will require many layers of annotation to give it meaning. The payoff from the reference work will come from understanding the proteins encoded by the genes.

Proteins not only make up the structural bulk of the human body but also include the enzymes that carry out the biochemical reactions of life. They are composed of units called amino acids linked together in a long string; each string folds in a way that determines the function of a protein. The order of the amino acids is set by the DNA base sequence of the gene that encodes a given protein, through intermediaries called RNA; genes that actively make RNA are said to be "expressed."

The Human Genome Project seeks not just to elucidate all the proteins produced within a human but also to comprehend how the genes that encode the proteins are expressed, how the DNA sequences of those genes stack up against comparable genes of other species, how genes vary within our species and how DNA sequences translate into observable characteristics. Layers of information built on top of the DNA sequence will reveal the knowledge embedded in the DNA. These data will fuel advances in biology for at least the next century. In a virtuous cycle, the more we learn, the more we will be able to extrapolate, hypothesize and understand.

By 2050 we believe that genomics will be able to answer the following major questions:

- *Will the three-dimensional structures of proteins be predictable from their amino acid sequences?*

The six billion bases of the human genome are thought to encode approximately 100,000 proteins. Although the sequence of amino acids in a protein can be translated in a simple step from the DNA sequence of a gene, we cannot currently elucidate the shape of a protein on purely theoretical grounds, and determining structures experimentally can be quite laborious. Still, a protein's structure is conserved—or maintained fairly constantly throughout evolution—much more than its amino acid sequence is. Many different amino acid sequences can

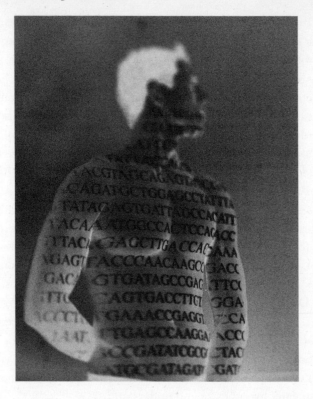

lead to proteins of similar shapes, so we can infer the structures of various proteins by studying a representative subset of proteins in detail.

Recently an international group of structural biologists have begun a Protein Structure Initiative to coordinate their work. Structural biologists "solve" the shapes of proteins either by making very pure crystals of a given protein and then bombarding the crystals with x-rays, or by subjecting a particular protein to nuclear magnetic resonance (NMR) analysis. Both techniques are time-consuming and expensive. The consortium intends to get the most information out of each new structure by using existing knowledge about related structures to group proteins into families that are most likely to share the same architectural features. Then the members of the consortium plan to target representatives of each family for examination by painstaking physical techniques.

As the catalogue of solved structures swells and scientists develop more refined schemes for grouping structures into a compendium of basic shapes, biochemists will increasingly be able to use computers to model the structures of newly discovered—or even wholly invented—proteins. Structural biologists project that a total of about 1,000 basic protein-folding motifs exist; current models suggest that solving just 3,000 to 5,000 selected structures, beyond the ones already known, could allow researchers to deduce the structures of new proteins routinely. With structural biologists solving more than 1,000 protein structures every year and with their progress accelerating, they should be able to complete the inventory not long after the human genome itself is sequenced.

- *Will synthetic life-forms be produced?*

Whereas structural biologists work to group proteins into categories for the practical aim of solving structures efficiently, the fact that proteins are so amenable to classification reverberates with biological meaning. It reflects how life on the earth evolved and opens the door to questions central to understanding the phenomenon of life itself. Is there a set of proteins common to all organisms? What are the biochemical processes required for life?

Already, with several fully sequenced genomes available— mostly from bacteria—scientists have started to take inventories of genes conserved among these organisms, guided by the grand question of what constitutes life, at least at the level of a single cell.

If, within a few years, investigators can expect to amass a tidy directory of the gene products—RNA as well as proteins—required for life, they may well be able to make a new organism from scratch by stringing DNA bases together into an

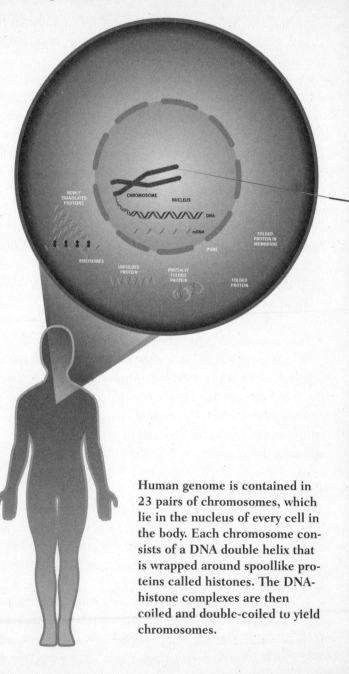

Human genome is contained in 23 pairs of chromosomes, which lie in the nucleus of every cell in the body. Each chromosome consists of a DNA double helix that is wrapped around spoollike proteins called histones. The DNA-histone complexes are then coiled and double-coiled to yield chromosomes.

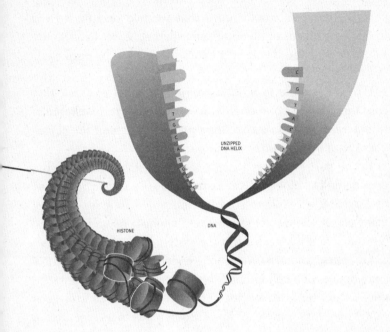

The ultimate aim of the Human Genome Project is to understand the proteins that are encoded by the DNA. When a gene
is "on," the cell uses a process called transcription to copy the
gene's DNA into a single-stranded molecule called messenger
RNA (mRNA), which leaves the nucleus to associate with a
series of large protein structures called ribosomes. The ribosomes then translate the mRNA into the chain of amino acids
that makes up the encoded protein. The new protein—here a
receptor destined for the cell membrane—goes through several
folding steps in a sequence that researchers are just beginning
to understand.

invented genome coding for invented products. If this invented genome crafts a cell around itself and the cell reproduces reliably, the exercise would prove that we had deciphered the basic mechanisms of life. Such an experiment would also raise safety, ethical and theological issues that cannot be neglected.

● *Will we be able to build a computer model of a cell that contains all the components, identifies all the biochemical interactions and makes accurate predictions about the consequences of any stimulus given to that cell?*

In the past 50 years, a single gene or a single protein often dominated a biologist's research. In the next 50 years, researchers will shift to studying integrated functions among many genes, the web of interactions among gene pathways and how outside influences affect the system.

Of course, biologists have long endeavored to describe how components of a cell interact: how molecules called transcription factors bind to specific scraps of DNA to control gene expression, for example, or how insulin binds to its receptor on the surface of a muscle cell and triggers a cascade of reactions in the cell that ultimately boosts the number of glucose transporters in the cell membrane. But the genome project will spark similar analyses for thousands of genes and cell components at a time. Within the next half-century, with all genes identified and all possible cellular interactions and reactions charted, pharmacologists developing a drug or toxicologists trying to predict whether a substance is poisonous may well turn to computer models of cells to answer their questions.

● *Will the details of how genes determine mammalian development become clear?*

Being able to model a single cell will be impressive, but to understand fully the life-forms we are most familiar with, we will plainly have to consider additional levels of complexity. We will have to examine how genes and their products behave in place and time—that is, in different parts of the body and in a body that changes over a life span. Developmental biologists have started to monitor how pools of gene products vary as tissues develop, in an attempt to find products that define stages of development. Now scientists are devising so-called expression arrays that survey thousands of gene products at a time, charting which ones turn on or off and which ones fluctuate in intensity of expression. Techniques such as these highlight many good candidates for genes that direct development and establish the animal body plan.

As in the past, model organisms—like the fruit fly *Drosophila*, the nematode *Caenorhabditis elegans* and the mouse—will remain the central workhorses in developmental biology. With the genome sequence of *C. elegans* and *Drosophila's* complete, the full human sequence on the way by 2003 and the mouse's likely within four to five years, sequence comparisons will become more commonplace and thorough and will give biologists many clues about where to look for the driving forces that fashion a whole animal. Many more complete genomes representing diverse branches of the evolutionary tree will be derived as the cost of sequencing decreases.

So far developmental biologists have striven to find signals that are universally important in establishing an animal's body plan, the arrangement of its limbs and organs. In time, they will also describe the variations—in gene sequence and perhaps in gene regulation—that generate the striking diversity of forms among different species. By comparing species, we will learn how genetic circuits have been modified to carry out distinct programs so that almost equivalent networks of genes fashion, for example, small furry legs in mice and arms with opposable digits in humans.

• *Will understanding the human genome transform preventive, diagnostic and therapeutic medicine?*

Molecular biology has long held out the promise of trans-

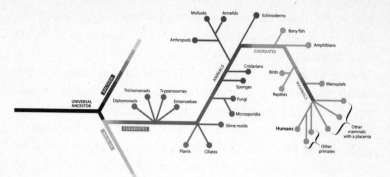

Tree of life illustrates the current view of the relationships among all living things, including humans. Once the DNA sequence of the human genome is known, scientists will be able to compare the information to that produced by efforts to sequence the genomes of other species, yielding a fuller understanding of how life on the earth evolved.

forming medicine from a matter of serendipity to a rational pursuit grounded in a fundamental understanding of the mechanisms of life. Its findings have begun to infiltrate the practice of medicine; genomics will hasten the advance. Within 50 years, we expect comprehensive genomics-based health care to be the norm in the U.S. We will understand the molecular foundation of diseases, be able to prevent them in many cases, and design accurate, individual therapies for illnesses.

In the next decade, genetic tests will routinely predict individual susceptibility to disease. One intention of the Human Genome Project is to identify common genetic variations. Once a list of variants is compiled, epidemiological studies will tease out how particular variations correlate with risk for disease. When the genome is completely open to us, such studies will reveal the roles of genes that contribute weakly to diseases on their own but that also interact with other genes and environmental influences such as diet, infection and prenatal exposure to affect health. By 2010 to 2020, gene therapy should also become a common treatment, at least for a small set of conditions.

Within 20 years, novel drugs will be available that derive from a detailed molecular understanding of common illnesses such as diabetes and high blood pressure. The drugs will target molecules logically and therefore be potent without significant side effects. Drugs such as those for cancer will routinely be matched to a patient's likely response, as predicted by molecular fingerprinting. Diagnoses of many conditions will be much more thorough and specific than they are now. For example, a patient who learns that he has high cholesterol will also know which genes are responsible, what effect the high cholesterol is likely to have, and what diet and pharmacological measures will work best for him.

By 2050 many potential diseases will be cured at the molecular level before they arise, although large inequities worldwide in access to these advances will continue to stir tensions. When people become sick, gene therapies and drug therapies will home in on individual genes, as they exist in individual people, making for precise, customized treatment. The average life span will reach 90 to 95 years, and a detailed understanding of human aging genes will spur efforts to expand the maximum length of human life.

• *Will we reconstruct accurately the history of human populations?*

Despite what may seem like great diversity in our species, studies from the past decade show that the human species is more homogeneous than many others; as a group, we display less variation than chimps do. Among humans, the same genetic variations tend to be found across all population groups, and only a small fraction of the total variation (between 10 and 15 percent) can be related to differences between groups. This has led some population biologists to the conclusion that not so long ago the human species was composed of a small group, perhaps 10,000 individuals, and that human populations dispersed over the earth only recently. Most genetic variation predated that time.

Armed with techniques for analyzing DNA, population geneticists have for the past 20 years been able to address anthropological questions with unprecedented clarity. Demographic events such as migrations, population bottlenecks and expansions alter gene frequencies, leaving a detailed and comprehensive record of events in human history. Genetic data have bolstered the view that modern humans originated relatively recently, perhaps 100,000 to 200,000 years ago, in Africa, and dispersed gradually into the rest of the world. Anthropologists have used DNA data to test cultural traditions about the origins of groups of humans, such as Gypsies and Jews, to track the migration into the South Pacific islands and the Americas, and to glean insights into the spread of populations in Europe, among other examples. As DNA sequence data become increasingly easy to accumulate, relationships among groups of people will become clearer, revealing histories of intermingling as well as periods of separation and migration. Race and ethnicity will prove to be largely social and cultural ideas; sharp, scientifically based boundaries between groups will be found to be nonexistent.

By 2050, then, we will know much more than we do now about human populations, but a question remains: How much can be known? Human beings have mated with enough abandon that probably no one family tree will be the unique solution accounting for all human history. In fact, the history of human populations will emerge not as a tree but as a trellis where lineages often meet and mingle after intervals of separation. Still, in 50 years, we will know how much ambiguity remains in our reconstructed history.

• *Will we be able to reconstruct the major steps in the evolution of life on the earth?*

Molecular sequences have been indispensable tools for drawing taxonomies since the 1960s. To a large extent, DNA sequence data have already exposed the record of 3.5 billion years of evolution, sorting living things into three domains— Archaea (single-celled organisms of ancient origin), Bacteria and Eukarya (organisms whose cells have a nucleus)—and revealing the branching patterns of hundreds of kingdoms and divisions. One aspect of inheritance has complicated the hope of assigning all living things to branches in a single tree of life. In many cases, different genes suggest different family histories for the same organisms; this reflects the fact that DNA isn't always inherited in the straightforward way, parent to offspring, with a more or less predictable rate of mutation marking the passage of time. Genes sometimes hop across large evolutionary gaps. Examples of this are mitochondria and chloroplasts, the energy-producing organelles of animals and plants, both of which contain their own genetic material and descended from bacteria that were evidently swallowed whole by eukaryotic cells.

This kind of "lateral gene transfer" appears to have been common enough in the history of life, so that comparing genes among species will not yield a single, universal family tree. As with human lineages, a more apt analogy for the history of life will be a net or a trellis, where separated lines diverge and join again, rather than a tree, where branches never merge.

In 50 years, we will fill in many details about the history of life, although we might not fully understand how the first self-replicating organism came about. We will learn when and how, for instance, various lineages invented, adopted or adapted genes to acquire new sets of biochemical reactions or different body plans. The gene-based perspective of life will have taken hold so deeply among scientists that the basic unit they consider will very likely no longer be an organism or a species but a gene. They will chart which genes have traveled together for how long in which genomes. Scientists will also address the question that has dogged people since Charles Darwin's day: What makes us human? What distinguishes us as a species?

Undoubtedly, many other questions will arise over the next 50 years as well. As in any fertile scientific field, the data will fuel new hypotheses. Paradoxically, as it grows in importance, genomics itself may not even be a common concept in 50 years, as it radiates into many other fields and ultimately becomes absorbed as part of the infrastructure of all biomedicine.

• *How will individuals, families and society respond to this explosion in knowledge about our genetic heritage?*

This social question, unlike the preceding scientific, technological and medical ones, does not come down to a yes-or-no answer. Genetic information and technology will afford great opportunities to improve health and to alleviate suffering. But any powerful technology comes with risks, and the more powerful the technology, the greater the risks. In the case of genetics, people of ill will today use genetic arguments to try to justify bigoted views about different racial and ethnic groups. As technology to analyze DNA has become increasingly widespread, insurers and employers have used the information to deny workers access to health care and jobs. How we will come to terms with the explosion of genetic information remains an open question.

Finally, will antitechnology movements be quieted by all the revelations of genetic science? Although we have enumerated so many questions to which we argue the answer will be yes, this is one where the answer will probably be no. The tension between scientific advances and the desire to return to a simple and more "natural" lifestyle will probably intensify as genomics seeps into more and more of our daily lives. The challenge will be to maintain a healthy balance and to shoulder collectively the responsibility for ensuring that the advances arising from genomics are not put to ill use.

Esteemed genomic researcher Willliam A. Haseltine, Ph.D., is the CEO of Human Genome Sciences, one of the first biotechnology companies to use new, cutting-edge genomics technologies to develop novel drugs. In the following essay, he breaks down the science behind employing the genetic code in the discovery and development of treatment and cures for diseases.

Discovering Genes for New Medicines

William A. Haseltine

Most readers are probably familiar with the idea of a gene as something that transmits inherited traits from one generation to the next. Less well appreciated is that malfunctioning genes are deeply involved in most diseases, not only inherited ones. Cancer, atherosclerosis, osteoporosis, arthritis and Alzheimer's disease, for example, are all characterized by specific changes in the activities of genes. Even infectious disease usually provokes the activation of identifiable genes in a patient's immune system. Moreover, accumulated damage to genes from a lifetime of exposure to ionizing radiation and injurious chemicals probably underlies some of the changes associated with aging.

A few years ago I, and some like-minded colleagues, decided that knowing where and when different genes are switched on in the human body would lead to far-reaching advances in our ability to predict, prevent, treat and cure disease. When a gene is active, or as a geneticist would say, "expressed," the sequence of the chemical units, or bases, in its DNA is used as a blueprint to produce a specific protein. Proteins direct, in various ways, all of a cell's functions. They serve as structural components, as catalysts that carry out the multiple chemical processes of life and as control elements that regulate cell reproduction, cell specialization and physiological activity at all levels. The development of a human from fertilized egg to mature adult is, in fact, the consequence of an orderly change in the pattern of gene expression in different tissues.

Knowing which genes are expressed in healthy and diseased tissues, we realized, would allow us to identify both the proteins required for normal functioning of tissues and the aberrations involved in disease. With that information in hand, it would be possible to develop new diagnostic tests for various illnesses and new drugs to alter the activity of affected proteins or genes. Investigators might also be able to use some of the proteins and genes we identified as therapeutic agents in their own right. We envisaged, in a sense, a high-resolution description of human anatomy descending to the molecular level of detail.

It was clear that identifying all the expressed genes in each of the dozens of tissues in the body would be a huge task. There are some 100,000 genes in a typical human cell. Only a small proportion of those genes (typically about 15,000) is expressed in any one type of cell, but the expressed genes vary from one cell type to another. So looking at just one or two cell types would not reveal the genes expressed in the rest of the body. We would also have to study tissues from all the stages of human development. Moreover, to identify the changes in gene expression that contribute to sickness, we would have to analyze diseased as well as healthy tissues.

Technological advances have provided a way to get the job done. Scientists can now rapidly discover which genes are expressed in any given tissue. Our strategy has proved the quickest way to identify genes of medical importance.

Take the example of atherosclerosis. In this common condi-

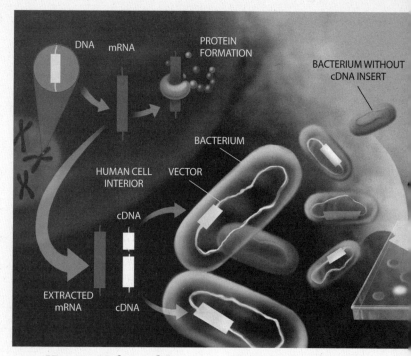

DNA   mRNA   PROTEIN FORMATION

BACTERIUM WITHOUT cDNA INSERT

BACTERIUM

HUMAN CELL INTERIOR   VECTOR

cDNA

EXTRACTED mRNA   cDNA

How to Make and Separate cDNA Molecules

Cells use messenger RNA to make protein. We discover genes by making complementary DNA (cDNA) copies of messenger RNA. First we have to clone and produce large numbers of copies of each cDNA, so there will be enough to determine its constituent bases. Molecular biologist have developed ways to insert cDNA into specialized DNA loops, called vectors, that can reproduce inside bacterial cells. A mixture of cDNAs from a given tissue is called a library.

Researchers at HGS have now prepared human cDNA libraries from almost all normal organs and tissues, as well as from many that are diseased. To make multiple copies of a library, we add it to bacteria that take up the vectors.

tion, a fatty substance called plaque accumulates inside arteries, notably those supplying the heart. Our strategy enables us to generate a list of genes expressed in normal arteries, along with a measure of the level of expression of each one. We can

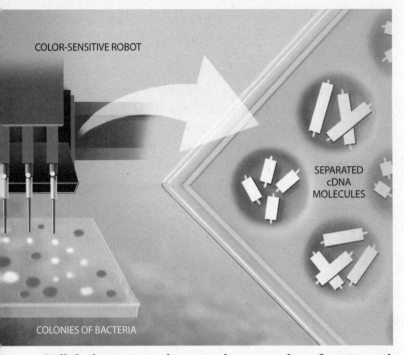

COLOR-SENSITIVE ROBOT

SEPARATED
cDNA
MOLECULES

COLONIES OF BACTERIA

All the bacteria are then spread out on a plate of nutrient gel and allowed to grow into colonies, so that each colony derives from a single bacterium. Next we use a robot that can automatically spot and pick off the gel of those colonies that did successfully acquire a cDNA. The robot accomplishes this by color. The vectors we use are designed so that if they fail to combine with a cDNA insert, they produce a blue pigment. The robot, which picks as many as 10,000 colonies of bacteria every day, identifies those containing human cDNA by avoiding blue ones. The cDNA from each picked colony, now in analyzable quantities, is then robotically purified.

then compare the list with one derived from patients with atherosclerosis. The difference between the lists corresponds to the genes (and thus the proteins) involved in the disease. It also indicates how much the genes' expression has been increased or decreased by the illness. Researchers can then make the human proteins specified by those genes.

Once a protein can be manufactured in a pure form, scientists can fairly easily fashion a test to detect it in a patient. A test to reveal overproduction of a protein found in plaque might expose early signs of atherosclerosis, when better options exist for treating it. In addition, pharmacologists can use pure proteins to help them find new drugs. A chemical that inhibited production of a protein found in plaque might be considered as a drug to treat atherosclerosis.

Our approach, which I call medical genomics, is somewhat outside the mainstream of research in human genetics. A great many scientists are involved in the Human Genome Project, an international collaboration devoted to the discovery of the complete sequence of the chemical bases in human DNA. (All the codes in DNA are constructed from an alphabet consisting of just four bases.) That information will be important for studies of gene action and evolution and will particularly benefit research on inherited diseases. Yet the genome project is not the fastest way to discover genes, because most of the bases that make up DNA actually lie outside genes. Nor will the project pinpoint which genes are involved in illness.

Genes by the Direct Route

Because the key to developing new medicines lies principally in the proteins produced by human genes, rather than the genes themselves, one might wonder why we bother with the genes at all. We could in principle analyze a cell's proteins directly. Knowing a protein's composition does not, however, allow us to make it, and to develop medicines, we must manufacture substantial amounts of proteins that seem important. The only practical way to do so is to isolate the corresponding genes and transplant them into cells that can express those genes in large amounts.

Our method for finding genes focuses on a critical intermediate product created in cells whenever a gene is expressed.

This intermediate product is called messenger RNA (mRNA); like DNA, it consists of sequences of four bases. When a cell makes mRNA from a gene, it essentially copies the sequence of DNA bases in the gene. The mRNA then serves as a template for constructing the specific protein encoded by the gene. The value of mRNA for research is that cells make it only when the corresponding gene is active. Yet the mRNA's base sequence, being simply related to the sequence of the gene itself, provides us with enough information to isolate the gene from the total mass of DNA in cells and to make its protein if we want to.

For our purposes, the problem with mRNA was that it can be difficult to handle. So we in fact work with a surrogate: stable DNA copies, called complementary DNAs (cDNAs) of the mRNA molecules. We make the cDNAs by simply reversing the process the cell uses to make mRNA from DNA.

The cDNA copies we produce this way are usually replicas of segments of mRNA rather than of the whole molecule, which can be many thousands of bases long. Indeed, different parts of a gene can give rise to cDNAs whose common origin may not be immediately apparent. Nevertheless, a cDNA containing just a few thousand bases still preserves its parent gene's unique signature. That is because it is vanishingly unlikely that two different genes would share an identical sequence thousands of bases long. Just as a random chapter taken from a book uniquely identifies the book, so a cDNA molecule uniquely identifies the gene that gave rise to it.

Once we have made a cDNA, we can copy it to produce as much as we want. That means we will have enough material for determining the order of its bases. Because we know the rules that cells use to turn DNA sequences into the sequences of amino acids that constitute proteins, the ordering of bases tells us the amino acid sequence of the corresponding protein fragment. That sequence, in turn, can be compared with the sequences in proteins whose structures are known. This maneu-

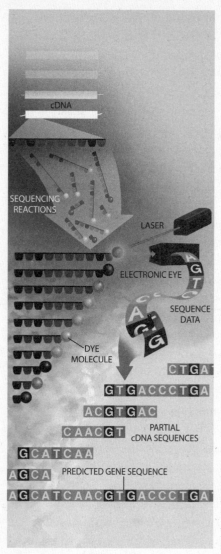

cDNA

SEQUENCING REACTIONS

LASER

ELECTRONIC EYE

SEQUENCE DATA

DYE MOLECULE

CTGA

GTGACCCTGA

ACGTGAC

CAACGT   PARTIAL cDNA SEQUENCES

GCATCAA

AGCA   PREDICTED GENE SEQUENCE

AGCATCAACGTGACCCTGA

How to Find a Partial cDNA Sequence

Researchers find partial cDNA sequences by chemically breaking down copies of a cDNA molecule to create an array of fragments that differ in length by one base. In this process, the base at one end of each fragment is attached to one of four florescent dyes, the color of the dye depending on the identity of the base in that position. Machines then sort the labeled fragments according to size. Finally, a laser excites the dye labels one by one. The result is a sequence of colors that can be read electronically and that corresponds to the order of the bases at one end of the cDNA being analyzed. Partial sequences hundreds of bases in length can be pieced together in a computer to produce complete gene sequences.

ver often tells us something about the function of the complete protein, because proteins containing similar sequences of amino acids often perform similar tasks.

Analyzing cDNA sequences used to be extremely time-consuming, but in recent years biomedical instruments have

normal vs → mRNA → cDNA → clone → fragment → sequence → compare to
diseased unique cDNA cDNA data
for bases
disease ie:
known
proteins

Discovering Genes for New Medicines / 47

been developed that can perform the task reliably and automatically. Another development was also necessary to make our strategy feasible. Sequencing equipment, when operated on the scale we were contemplating, produces gargantuan amounts of data. Happily, computer systems capable of handling the resulting megabytes are now available, and we and others have written software that helps us make sense of this wealth of genetic detail.

Assembling the Puzzle

Our technique for identifying the genes used by a cell is to analyze a sequence of 300 to 500 bases at one end of each cDNA molecule. These partial cDNA sequences act as markers for genes and are sometimes referred to as <u>expressed sequence tags.</u> We have chosen this length for our partial cDNA sequences because it is short enough to analyze fairly quickly but still long enough to identify a gene unambiguously. If a cDNA molecule is like a chapter from a book, a partial sequence is like the first page of the chapter—it can identify the book and even give us an idea what the book is about. Partial cDNA sequences, likewise, can tell us something about the gene they derive from. At HGS, we produce about a million bases of raw sequence data every day.

Our method is proving successful: we have identified thousands of genes, many of which may play a part in illness. Other companies and academic researchers have also initiated programs to generate partial cDNA sequences.

HGS's computers recognize many of the partial sequences we produce as deriving either from one of the genes researchers have already identified by other means or from a gene we have previously found ourselves. When we cannot definitely assign a newly generated partial sequence to a known gene, things get more interesting. Our computers then scan through our databases as well as public databases to see

whether the new partial sequence overlaps something some-
one has logged before. When we find a clear overlap, we piece together the overlapping partial sequences into ever lengthening segments
called <u>contigs</u>. Contigs correspond, then, to incomplete
sequences we infer to be present somewhere in a parent gene.
This process is somewhat analogous to fishing out the phrases
"a midnight dreary, while I pondered" and "while I pondered,
weak and weary/Over many a. . . volume" and combining them
into a fragment recognizable as part of Edgar Allan Poe's "The
Raven."

At the same time, we attempt to deduce the likely function
of the protein corresponding to the partial sequence. Once we
have predicted the protein's structure, we classify it according
to its similarity to the structures of known proteins. Sometimes
we find a match with another human protein, but often we
notice a match with one from a bacterium, fungus, plant or
insect: other organisms produce many proteins similar in function to those of humans. Our computers continually update
these provisional classifications.

In 1994, for example, we predicted that genes containing
four specific contigs would each produce proteins similar to
those known to correct mutations in the DNA of bacteria and
yeast. Because researchers had learned that failure to repair
mutations can cause colon cancer, we started to work out the
full sequences of the four genes. When a prominent colon cancer researcher later approached us for help in identifying genes
that might cause that illness—he already knew about one such
gene—we were able to tell him that we were already working
with three additional genes that might be involved.

Subsequent research has confirmed that mutations in any
one of the four genes can cause life-threatening colon, ovarian
or endometrial cancer. As many as one in every 200 people in
North America and Europe carry a mutation in one of these
mismatch repair genes, as they are called. Knowing this, scien-

tists can develop tests to assess the mismatch repair genes in people who have relatives with these cancers. If the people who are tested display a genetic pre-disposition to illness, they can be monitored closely. Prompt detection of tumors can lead to lifesaving surgery, and such tests have already been used in clinical research to identify people at risk.

Our database now contains more than a million cDNA-derived partial gene sequences, sorted into 170,000 contigs. Overall, more than half of the new genes we identify have a resemblance to known genes that have been assigned a probable function. As time goes by, this proportion is likely to increase.

If a tissue gives rise to an unusually large number of cDNA sequences that derive from the same gene, it provides an indication that the gene in question is producing copious amounts of mRNA. That generally happens when the cells are producing large amounts of the corresponding protein, suggesting that the protein may be doing a particularly vital job. HGS also pays particular attention to genes that are expressed only in a narrow range of tissues, because such genes are most likely to be useful for intervening in diseases affecting those tissues. Of the many genes we have discovered, we have identified a percentage that seem especially likely to be medically important.

New Genes, New Medicines

Using the partial cDNA sequence technique for gene discovery, researchers have for the first time been able to assess how many genes are devoted to each of the main cellular functions, such as defense, metabolism and so on. The vast store of unique information from partial cDNA sequences offers new possibilities for medical science. These opportunities are now being systematically explored.

Databases such as ours have already proved their value for finding proteins that are useful as signposts of disease. Prostate cancer is one example. A widely used test for detecting prostate cancer measures levels in the blood of a protein called prostate specific antigen. Patients who have prostate cancer often exhibit unusually high levels. Unfortunately, slow-growing, relatively benign tumors as well as malignant tumors requiring aggressive therapy can cause elevated levels of the antigen, and so the test is ambiguous.

HGS and its partners have analyzed mRNAs from multiple samples of healthy prostate tissue as well as from benign and malignant prostate tumors. We found about 300 genes that are expressed in the prostate but in no other tissue; of these, about 100 are active only in prostate tumors, and about 20 are expressed only in tumors rated by pathologists as malignant. We and our commercial partners are using these 20 genes and their protein products to devise tests to identify malignant prostate disease. We have similar work under way for breast, lung, liver and brain cancers.

Databases of partial cDNA sequences can also help find genes responsible for rare diseases. Researchers have long known, for example, that a certain form of blindness in children is the result of an inherited defect in the chemical breakdown of the sugar galactose. A search of our database revealed two previously unknown human genes whose corresponding proteins were predicted to be structurally similar to known galactose-metabolizing enzymes in yeast and bacteria. Investigators quickly confirmed that inherited defects in either of these two genes cause this type of blindness. In the future, the enzymes or the genes themselves might be used to prevent the affliction.

Partial cDNA sequences are also establishing an impressive record for helping researchers to find smaller molecules that are candidates to be new treatments. Methods for creating and testing small-molecule drugs—the most common type—have improved dramatically in the past few years. Automated equipment can rapidly screen natural and synthetic compounds for their ability to affect a human protein involved in disease, but the limited number of known protein targets has delayed progress. As more human proteins are investigated, progress should accelerate. Our work is now providing more than half of SmithKline Beecham's leads for potential products.

Databases such as ours not only make it easier to screen molecules randomly for useful activity. Knowing a protein's structure enables scientists to custom-design drugs to interact in a specific way with the protein. This technique, known as rational drug design, was used to create some of the new protease inhibitors that are proving effective against HIV (although our database was not involved in this particular effort). We are confident that partial cDNA sequences will allow pharmacologists to make more use of rational drug design.

One example of how our database has already proved useful concerns cells known as osteoclasts, which are normally present in bone; these cells produce an enzyme capable of degrading bone tissue. The enzyme appears to be produced in excess in some disease states, such as osteoarthritis and osteoporosis. We found in our computers a sequence for a gene expressed in osteoclasts that appeared to code for the destructive enzyme; its sequence was similar to that of a gene known to give rise to an enzyme that degrades cartilage. We confirmed that the osteoclast gene was responsible for the degradative enzyme and also showed that it is not expressed in other tissues. Those discoveries meant we could invent ways to thwart the gene's protein without worrying that the methods would harm other tissues. We then made the protein, and SmithKline Beecham has used it to identify possible therapies by a combination of high-through-put screening and rational drug design. The company has also used our database to screen for molecules that might be used to treat atherosclerosis.

One extremely rich lode of genes and proteins, from a medical point of view, is a class known as G-protein coupled receptors. These proteins span the cell's outer membrane and convey biological signals from other cells into the cell's interior. It is likely that drugs able to inhibit such vital receptors could be used to treat diseases as diverse as hypertension, ulcers, migraine, asthma, the common cold and psychiatric disorders. HGS has found more than 70 new G-protein coupled receptors. We are now testing their effects by introducing receptor genes we have discovered into cells and evaluating how the cells that make the encoded proteins respond to various stimuli. Two genes that are of special interest produce proteins that seem to be critically involved in hypertension and in adult-onset diabetes. Our partners in the pharmaceutical industry are searching for small molecules that should inhibit the biological signals transmitted by these receptors.

Last but not least, our research supports our belief that some of the human genes and proteins we are now discovering will, perhaps in modified form, themselves constitute new therapics. Many human proteins are already used as drugs; insulin and clotting factor for hemophiliacs are well-known examples. Proteins that stimulate the production of blood cells are also used to speed patients' recovery from chemotherapy.

The proteins of some 200 of the full-length gene sequences HGS has uncovered have possible applications as medicines. We have made most of these proteins and have instituted tests of their activity on cells. Some of them are also proving promising in tests using experimental animals. The proteins include several chemokines, molecules that stimulate immune system cells.

Developing pharmaceuticals will never be a quick process, because medicines, whether proteins, genes or small molecules, have to be extensively tested. Nevertheless, partial cDNA sequences can speed the discovery of candidate therapies. HGS allows academic researchers access to much of its database, although we ask for an agreement to share royalties from any ensuing products.

The systematic use of automated and computerized methods of gene discovery has yielded, for the first time, a comprehensive picture of where different genes are expressed—the anatomy of human gene expression. In addition, we are starting to learn about the changes in gene expression in disease. It is too early to know exactly when physicians will first successfully use this knowledge to treat disease. Our analyses predict, however, that a number of the resulting therapies will form mainstays of 21st-century medicine.

This essay is not just about the human genome but about the future of all of science, with the Human Genome Project placed within this larger context. Here Sir John Maddox, lecturer in theoretical physics at the University of Manchester from 1949 to 1956 but much better known as the editor-in-chief of the most prestigious British science journal, Nature, *from 1966 to 1973 and from 1980 to 1995, examines how science may progress during the first 50 years of the third millennium.*

Scientific breakthroughs occur more or less at random; we cannot predict when such an event might happen. They frequently occur, however, when a knowledge base is readily available. Automated genome sequencing is the breakthrough that made the Human Genome Project possible, but we had no reason to expect it until our computer technology and our knowledge of genetics had attained the "breakthrough level": the point where not only was such a breakthrough to be expected, it was almost inevitable. As the HGP continues, we will have an enormous and expanding knowledge base of the human genome, so we may expect correspondingly significant breakthroughs in several areas of medicine, biochemistry, and genetics. This expectation of new science is the real motivation that drives the Human Genome Project forward.

The Unexpected
Science to Come

Sir John Maddox

The questions we do not yet have the wit to ask will be a growing preoccupation of science in the next 50 years. That is what the record shows. Consider the state of science more than a century ago, in 1899. Then, as now, people were reflecting on the achievements of the previous 100 years. One solid success was the proof by John Dalton in 1808 that matter consists of atoms. Another was the demonstration (by James Prescott Joule in 1851) that energy is indeed conserved and the earlier surmise (by French physicist Sadi Carnot) that the efficiency with which one form of energy can be converted into another is inherently limited: jointly, those developments gave us what is called thermodynamics and the idea that the most fundamental laws of nature incorporate an "arrow of time."

There was also Charles Darwin, whose *Origin of Species by Means of Natural Selection* (published in 1859) purported to account for the diversity of life on Earth but said nothing about the mechanism of inheritance or even about the reasons why different but related species are usually mutually infertile.

Finally, in the 19th century's catalogue of self-contentment, was James Clerk Maxwell's demonstration of how electricity and magnetism can be unified by a set of mathematical equations on strictly Newtonian lines. More generally, Newton's laws had been so well honed by practice that they offered a solution for any problem in the real world that could be accurately defined. What a marvelous century the 1800s must have been!

Only the most perceptive people appreciated, in 1899, that there were flaws in that position. One of those was Hendrik Antoon Lorentz of Leiden University in the Netherlands, who saw that Maxwell's theory implicitly embodied a contradiction: the theory supposed that there must be an all-pervading ether through which electromagnetic disturbances are propagated, but it is far simpler to suppose that time passes more slowly on an object moving relative to an observer. It was a small step from there (via Henri Poincaré of the University of Paris) to Albert Einstein's special theory of relativity, published in 1905. The special theory, which implies that relative velocities cannot exceed the speed of light, falsifies Newton only philosophically: neither space nor time can provide a kind of invisible grid against which the position of an object, or the time at which it attains that position, can be measured. A century ago few people seem to have appreciated that A. A. Michelson and E. W. Morley, in the 1880s, had conducted an experiment whose simplest interpretation is that Maxwell's ether does not exist.

For those disaffected from, or even offended by, the prevailing complacency of 1899, ample other evidence should have hinted that accepted fundamental science was heading into trouble. Atoms were supposed to be indivisible, so how could one explain what seemed to be fragments of atoms, the electrons and the "rays" given off by radioactive atoms, discovered in 1897? Similarly, although Darwin had supposed that the inheritable (we would now say "genetic") changes in the constitution of individuals are invariably small ones, the rediscovery of Gregor Mendel's work in the 1850s (chiefly by Hugo de Vries in the Netherlands) suggested that spontaneous genetic changes are, rather, discrete and substantial. That development led, under the leadership of Thomas Hunt Morgan, to the emergence of Columbia University in New York City as the citadel of what is now called classical genetics (a phrase coined only in 1906) and to the recognition in the 1930s that the contradiction between Darwinism and "Mendel-Morganism" (as the Soviets in the 1950s came to call Columbia's work) is not as sharp as it first seemed.

Now we marvel at how these contradictions have been resolved and at much else. Our own contentment with our own century surpasses that of 1899. Not least important is the sense of personal liberation we enjoy that stems from applications of science in the earliest years of the 20th century—Marconi's bridging of the Atlantic with radio waves and the Wright brothers' measured mile of flight in a heavier-than-air machine. (Wilbur and Orville had built a primitive wind tunnel at their base in Ohio before risking themselves aloft.) The communications and aviation industries have grown from those beginnings. Our desks are cluttered with powerful computing machines that nobody foresaw in 1900. And we are also much healthier: think of penicillin!

A Catalogue of Contentment

In fundamental science, we have as much as or more to boast about than did the 19th century. Special relativity is no longer merely Newton made philosophically respectable. Through its implication that space and time must be dealt with on an equal footing, it has become a crucial touchstone of the validity of theories in fundamental physics.

The other three landmarks in fundamental science this century were hardly foreseen. Einstein's general theory of relativity in 1915, which would have been better called his "relativistic theory of gravitation," would have been a surprise to all but close readers of Ernst Mach, the Viennese physicist and positivist philosopher. By positing that gravitational forces everywhere are a consequence of a gravitational field that reaches into the farthest corners of the cosmos, Einstein launched the notion that the structure and evolution of the universe are ineluctably linked. But even Einstein was surprised when Edwin Hubble discovered in 1929 that the universe is expanding.

Quantum mechanics was another bolt from the blue, even though people had been worrying about the properties of the radiation from hot objects for almost half a century. The problem was to explain how it arises that the radiation from an object depends crucially on its temperature such that the most prominent frequency in the emission is directly proportional to the temperature, at least when the temperature is measured from the absolute zero (which is 273 degrees Celsius below the freezing point of water, or -459 degrees Fahrenheit, and which had itself been defined by 19th-century thermodynamics). The solution offered by Max Planck in 1900 was that energy is transferred between a hot object and its surroundings only in finite (but very small) amounts, called quanta. The actual amount of energy in a quantum depends on the frequency of the radiation and, indeed, is proportional to it. Planck confessed at the time that he did not know what this result meant and guessed that his contemporaries would also be perplexed.

As we know, it took a quarter of a century for Planck's difficulty to be resolved, thanks to the efforts of Niels Bohr, Werner Heisenberg, Erwin Schrödinger and Paul Dirac, together with a small army of this century's brightest and best. Who would have guessed, in 1900, that the outcome of the enterprise Planck began would be a new system of mechanics, as comprehensive as Newton's in the sense that it is applicable to all well-posed problems but applies only to atoms, molecules and the parts thereof—electrons and so on?

Even now there are people who claim that quantum mechanics is full of paradoxes, but that is a deliberate (and often mischievous) reading of what happened in the first quarter of this century. Our intuitive understanding of how objects in the macroscopic world behave (embodied in Newton's laws) is based on the perceptions of our senses, which are themselves the evolutionary products of natural selection in a world in which the avoidance of macroscopic objects (predators) or their capture (food) would have favored survival of the species. It is difficult to imagine what selective advantage our ancestors would have gained from a capacity to sense the behavior of subatomic particles. Quantum mechanics is therefore not a paradox but rather a discovery about the nature of reality on scales (of time and distance) that are very small. From that revelation has flowed our present understanding of how particles of nuclear matter may be held to consist of quarks and the like—an outstanding intellectual achievement, however provisional it may be.

The third surprise this century has followed from the discovery of the structure of DNA by James D. Watson and Francis Crick in 1953. That is not to suggest that Watson and Crick were unaware of the importance of their discovery. By the early 1950s it had become an embarrassment that the genes, which the Columbia school of genetics had shown are arranged in a linear fashion along the chromosomes, had not been assigned a chemical structure of some kind. The surprise was that the structure of DNA accounted not just for how offspring inherit their physical characteristics from their parents but also for how individual cells in all organisms survive from millisecond to millisecond in the manner in which natural selection has shaped them. The secret of life is no longer hidden.

A Catalogue of Ignorance

Both quantum mechanics and the structure of DNA have

enlarged our understanding of the world to a degree that their originators did not and could not have foretold. There is no way of telling which small stone overturned in the next 50 years will lead to a whole new world of science. The best that one can do is make a catalogue of our present ignorance—of which there is a great deal—and then extrapolate into the future current trends in research. Yet even that procedure suggests an agenda for science in the next half a century that matches in its interest and excitement all that has happened in the century now at an end. Our children and grandchildren will be spellbound.

One prize now almost ready for the taking is the reconstruction of the genetic history of the human race, *Homo sapiens*. A triumph of the past decade has been the unraveling of the genetics of ontogeny, the transformation of a fertilized embryo into an adult in the course of gestation and infancy. The body plans of animals and plants appear initially to be shaped by genes of a common family (called *Hox* genes) and then by species-specific developmental genes. Although molecular biologists are still struggling to understand how the hierarchical sequence of developmental genes is regulated and how genes that have done their work are then made inactive, it is only a matter of time before the genes involved in the successive stages of human development are listed in the order in which they come into play.

Then it will be possible to tell from a comparison between human and, say, chimpanzee genes when and in what manner the crucial differences between humans and the great apes came into being. The essence of the tale is known from the fossil record: the hominid cerebral cortex has steadily increased in size over the past 4.5 million years; hominids were able to walk erect with *Homo erectus* 2.1 million years ago; and the faculty of speech probably appeared with mitochondrial Eve perhaps as recently as 125,000 years ago. Knowing the genetic basis of these changes will give us a more authentic history of our species and a deeper understanding of our place in nature.

That understanding will bring momentous by-products. It may be possible to infer why some species of hominids, of which the Neanderthals are only one, failed to survive to modern times. More important is that the genetic history of *H. sapiens* is likely to be a test case for the mechanism of speciation. Despite the phrase "Origin of Species" in the title of Darwin's great book, the author had nothing to say about why members of different species are usually mutually infertile. Yet the most striking genetic difference between humans and the great apes is that humans have 46 chromosomes (23 pairs), whereas our nearest relatives have 48. (Much of the missing ape chromosome seems to be at the long end of human chromosome 2, but other fragments appear elsewhere in the human genome, notably on the X chromosome.) It will be important for biology generally to know whether this rearrangement of the chromosomes was the prime cause of human evolution or whether it is merely a secondary consequence of genetic mutation.

The 50 years ahead will also see an intensification of current efforts to identify the genetic correlates of evolution more generally. Comparison of the amino acid sequences of similar proteins from related species or of the sequences of nucleotides in related nucleic acids—the RNA molecules in ribosomes are a favorite—is in principle a way of telling the age of the common ancestor of the two species. It is simply necessary to know the rate at which mutations naturally occur in the molecules concerned.

But that is not a simple issue. Mutation rates differ from one protein or nucleic acid molecule to another and vary from place to place along their length. Constructing a more reliable "molecular clock" must be a goal for the near future. (The task is similar to, but if anything more daunting than, cosmologists' effort to build a reliable distance-scale for the universe.) Then we shall be able to guess at the causes of the great turning points in the evolution of life on Earth—the evolution of the Krebs cycle by which all but bacterial cells turn chemicals into energy, the origin of photosynthesis, the appearance of the first multicellular organisms (now firmly placed more than 2,500 million years ago).

With luck, the same effort will also tell us something about the role of viruslike agents in the early evolution of life. The human genome is crammed with DNA sequences that appear to be nucleic acid fossils of a time when genetic information was readily transferred between different species much as bacteria in the modern world acquire certain traits (such as resistance to antibiotics) by exchanging DNA structures called plasmids. We shall not know our true place in nature until we understand how the apparently useless DNA in the human genome (which Crick was the first to call "junk") contributed to our evolution.

Understanding all the genomes whose complete structure is known will not, in itself, point back to the origin of life as such. It should, however, throw more light on the nature of living things in the so-called RNA world that is supposed to have preceded the predominantly DNA life that surrounds us. It is striking and surely significant of something that modern cells still use RNA molecules for certain basic functions—as the editors of DNA in the nucleus, for example, and as the templates for making the structures called telomeres that stabilize the ends of chromosomes.

At some stage, but probably more than half a century from now, someone will make a serious attempt to build an organism based on RNA in the laboratory. But the problem of the origin of life from inorganic chemicals needs understanding now lacking—not least an understanding of how flux of radiation such as that from the sun can, over time, force the formation of complex from simpler chemicals. Something of the kind is known to occur in giant molecular clouds within our galaxy, where radioastronomers have been finding increasingly complex chemicals, most recently fullerenes (commonly called "bucky-balls") such as C60. The need is for an understanding of the relation between complexity and the flux of radiation. This is a problem in irreversible thermodynamics to which too little attention has been paid.

Indeed, biologists in general have paid too little attention to the quantitative aspects of their work in the past few hectic decades. That is understandable when there are so many interesting (and important) data to be gathered. But we are already at the point where deeper understanding of how, say, cells function is impeded by the simplification of reality now commonplace in cell biology and genetics—and by the torrent of data accumulating everywhere. Simplification? In genetics, it is customary to look for (and to speak of) the "function" of a newly discovered gene. But what if most of the genes in the human genome, or at least their protein products, have more than one function, perhaps even mutually antagonistic ones? Plain-language accounts of cellular events are then likely to be misleading or meaningless unless backed up by quantitative models of some kind.

A horrendous example is the cell-division cycle, in which the number of enzymes known to be involved seems to have been growing for the past few years at the rate of one enzyme a week. It is a considerable success that a complex of proteins that functions as a trigger for cell division (at least in yeast) has been identified, but why this complex functions as a trigger and how the trigger itself is triggered by influences inside and outside a cell are questions still unanswered. They will remain so until researchers have built numerical models of cells in their entirety. That statement is not so much a forecast as a wish.

Despite the illusion we enjoy that the pace of discovery is accelerating, it is important that, in some fields of science, many goals appear to be attainable only slowly and by huge collective effort. To be sure, the spacecrafts now exploring the solar system are usually designed a decade or so before their launch. After a century of seismology, only now are measurement and analytical techniques sensitive enough to promise that we shall soon have a picture of the interior of the planet on which we live, one that shows the rising convection plumes of mantle rock that drive the tectonic plates across the surface of Earth. Since the 1960s, molecular biologists have had the goal of understanding the way in which the genes of living organisms are regulated, but not even the simplest bacterium has yet been comprehensively accounted for. And we shall be lucky if the neural correlates of thinking are identified in the half-century ahead. The application of what we know already will enliven the decades immediately ahead, but there are many important questions that will be answered only with great difficulty.

And we shall be surprised. The discovery of living things of some kind elsewhere in the galaxy would radically change the general opinion of our place in nature, but there will be more subtle surprises, which, of necessity, cannot be anticipated. They are the means by which the record of the past 500 years of science has been repeatedly enlivened. They are also the means by which the half-century ahead will enthrall the practitioners and change the lives of the rest of us.

While the completion of the human genome sequencing in June 2000 was remarkable, it was the tip of the iceberg. "Beyond the First Draft" details the larger job of annotating the genes, which is comprised of determining each genes role and how it interacts with other genes. The essay profiles the companies jockeying to speed up the annotation process through universal programs and accessible databases.

Beyond the First Draft

Tabitha M. Powell

Unprecedented fanfare greeted the June 26, 2000 announcement that scientists had completed a draft of the human genome sequence. The truth is, however, that figuring out the order of the letters in our genetic alphabet was the easy part. Now comes the hard part: deciphering the meaning of the genetic instruction book.

The next stage goes by a deceptively prosaic name: annotation. Strictly speaking, "annotation" comprises everything that can be known about a gene: where it works, what it does and how it interacts with fellow genes. Right now, scientists often use the term simply to signify the first step: gene finding. That means discovering which parts of a stretch of DNA belong to a gene and distinguishing them from the other 96 percent or so that have no known function, often called junk DNA.

Several companies have sprouted up to provide bioinformatics tools, software and services. Their success, though, may hinge on a peaceful spot south of England's University of Cambridge. It is home to the Sanger Center, the U.K. partner in the publicly funded Human Genome Project (HGP) consortium, and the European Bioinformatics Institute (EBI), Europe's equivalent of the National Center for Biotechnology Information (NCBI) at the National Institutes of Health. Sanger and EBI are collaborating on the Ensemble project, which consists of computer programs for genome analysis and the public database of human DNA sequences. New DNA sequences arrive in bits and pieces; automated routines scan the sequences, looking for patterns typically found in genes. "One of the important things about Ensembl is that we're completely open, so you can see all our data, absolutely everything," says EBI's Ewan Birney.

No matter how talented their algorithms, however, computers can't get all the genes, and they can't get them all right. Many additions and corrections, plus the all-important information about how genes are regulated and what they do, are tasks for human curators. That problem may be solved for Ensembl by a distributed computing system under development by Lincoln Stein of the Cold Spring Harbor Laboratory on Long Island, N.Y. The plan is to provide human annotation—corrections and suggestions and research findings from scientists around the world—layered on top of Ensembl's automatic annotation. Stein's Distributed Sequence Annotation System, DAS for short, borrows an approach from Napster, the controversial software that allows people to swap music files over the Internet.

The plan is that different labs will publish their own annotations (on dedicated servers) according to specifications of some commonly accepted map of the genome—like Ensembl's. "Then the browser application would be able to go out onto the Web, find out what's there and bring it all into an integrated view so that you could see in a graphical way what different people had to say about a region of the genome," Stein explains. In this way DAS may solve a huge problem that plagues biology databases: the lack of a standard format for archiving and presenting data, which, among other disadvantages, makes it impossible to search across them and compare contents.

The DAS model is not universally beloved. NCBI director David Lipman is concerned that the human annotations may be full of rubbish because they will not be peer-reviewed. Stein acknowledges the possibility but hopes that good annotation will drive out bad. He is more concerned about whether the spirit of volunteerism will flag when faced with personnel changes and the vagaries of funding. Keeping a lab's Web server running and up-to-date is a long-term commitment.

As opposed to the well-publicized rivalry between the HGP and the privately owned Celera Genomics in sequencing the genome, many bioinformatics firms don't regard Ensembl as an organization to beat. In fact, several commercial players endorse collaboration; financial opportunity will come from using the data in a unique way. James I. Garrels, president of Proteome in Beverly, Mass., expects to partner with and provide help to public-domain efforts to amass a basic description of each gene, its protein and a few of the protein's key properties. But Proteome also believes that nothing beats the vast and versatile human brain for making sense of the vast and versatile human genome. The company's researchers scour the literature, concentrating on proteins—the product most genes make—and since 1995 have built protein data-bases on three model organisms: the roundworm *Caenorhabditis elegans* and two species of yeast. Now they are adding data on the human, mouse and rat genomes. The company's niche will be integrating all that information. "That's not the type of effort contemplated in the public domain," Garrels points out.

Proteome's strength is likely to lie in its customers' ability to compare sequences across species. Because evolution has conserved a great many genes and used them over and over, such comparisons are a rich source of hints: a human gene whose job is currently a mystery will often be nearly identical to one present in other species.

Randy Scott, president of Incyte Genomics in Palo Alto, Calif., is another fan of sharing the load. Besides, "there's plenty of ways to make money," Scott declares. "We assume there are going to be broadly annotated databases available in the public domain, and the sooner we can get there, the faster Incyte can focus on down-stream, on how we take that information to create new levels of information." For instance, the company has picked a group of genes it believes will be important for diagnostics and other applications and is concentrating its annotation efforts on them. It also has databases that permit some cross-species comparisons.

Given Ensembl's open-source code, distributed annotation and determination to stay free, comparisons to the free Linux computer operating system—which may someday challenge Microsoft Windows's supremacy—are natural. But the parallel doesn't go very far. Thinking of public and commercial annotation products as rivals misses the point, observers say. In the words of Sean Eddy of Washington University, who is working on DAS: "The human genome is too big for anybody to look at alone. We're going to have to figure out ways for the public and private sectors to work collaboratively rather than competitively."

Expanding on what still needs to occur, the following essay details the steps needed to be completed before scientists truly understand the human genome. The steps range from proofreading the original genetic sequencing to the more difficult process of identifying gene functions.

Genome Scientists' To-Do List

Tabitha M. Powell

1. Correct errors and proofread. The original plan was to repeat the sequencing up to 12 times to prune away the mistakes that inevitably accompany a project involving 3.1 billion pieces of datum. In the rush to make the joint announcement, the privately funded Celera Genomics and the publicly funded international consortium Human Genome Project settled temporarily for less than half that goal. Proofreading will probably take another year or two from the date of the announcement.

2. Fill tens of thousands of gaps in the sequence. These holes amounted to about 15 percent of the genome on June 26, 2000. Most gaps lie in stretches of short sequences repeated hundreds or thousands of times, which makes them enormously difficult to get right.

3. Sequence the seven percent of the human genome that was originally excluded by design. This region is heterochromatin, highly condensed DNA long believed to contain no genes. But in March 2000, analysis revealed that fruit fly heterochromatin (about one third of the fly's genome)

appears to contain about 50 genes—so human heterochromatin probably contains a few genes, too.

4. Finish finding all the genes that make proteins. This step takes place after the sequence is cleaned up and deemed 99.99 percent accurate. About 38,000 protein-coding genes have been confirmed so far. Recent estimates have tended to fall below 60,000.

5. Find the non-protein-making genes. There are, for instance, genes that make RNA rather than protein. They tend to fall below the threshold of today's gene-finding software, so new ways of discovering them will have to be devised.

6. Discover the regulatory sequences that activate a gene and that govern how much of its product to make.

7. Untangle the genes' intricate interactions with other molecules.

8. Identify gene functions. Because a gene may make several proteins, and each protein may perform more than one job, the task will be stupendous.

As they check each item off the list, researchers will be generating the information that will make it possible to attack and even prevent a vast array of human ills. But how long will it take to get through the checklist? If anyone knows, it should be Celera president J. Craig Venter. On announcement day Venter predicted that the analysis will take most of this century.

> *To speed up the process of finding each gene's function, the following essay details a new technique developed by the Whitehead Institute at the Massachusetts Institute of Technology and Corning, Inc. This technique, called the DNA microarray technique, identifies within in a week which cellular circuits are controlled by which master switches in the genome.*

e, all new drugs must ultima
sted in mammals—and that of
ce. Mice are very close to
rms of their genome: more th
rcent of the mouse proteins
far similar enough to kr
oteins. Ten laboratories ad
called the Mouse Genome Se
twork, collectively received
om the National Institutes

Finding the Genome's Master Switches

Kristin Leutwyler

O nce the Human Genome Project delivers a list of all of our genes, the next trick will be figuring out just what those genes do. Making the job a little easier, though, is a new DNA microarray technique developed by scientists at the Whitehead Institute at the Massachusetts Institute of Technology and Corning, Inc. Reporting in the December 2000 issue of *Science*, Richard Young and his colleagues unveil a method that identifies in about a week which cellular circuits are controlled by which master switches in the genome—a task that ordinarily takes years. "We are very excited by these results because they suggest that our technique can be used to create a 'user's manual' for the cell's master controls, a booklet that matches the master switches to the circuits they control in the genome," Young says.

The master switches are in fact proteins, called gene activators, that bind to specific regions of DNA, or genes, and in doing so, initiate series of steps that control everything from cell growth and development to seeding disease. Young's group built their method around DNA microarrays because these devices make it possible to take a kind of snapshot of a cell, and see which genes are turned on and which are turned off. For biologists, knowing which genes are active as a cell performs some function is incredibly useful information, much like being able to match the individual notes in a chord with the sound they produce. But it doesn't reveal which master switch—or hand—played the notes, and the human genome contains about 1,000 master switches. To date, scientists know the activity of only a quarter of them, such as the p53 protein, which plays a role in cancer.

The first step in the new technique is fixing the master switch proteins in living cells to their DNA binding sites using chemical crosslinking methods—sort of like gluing the hands to the keys they are striking. The scientists then open the cells, creating a soup of protein-DNA complexes. They use antibodies with magnetic beads to draw out interesting fragments of DNA, with the master switch protein still attached. Next they isolate the DNA fragments, label them with fluorescent dye and hybridize them to a DNA array containing genomic DNA from yeast, which identifies what they are. As a test of the method, Young and his colleagues demonstrated that it can successfully pick out the cellular circuits controlled by two known master switches. "Our goal is to use this technique to find the circuits controlled by the 200 or so master switches in yeast," Young adds, "and then develop analogous techniques in humans."

Using information from the human genome, companies such as Myriad Genetics are developing more accurate, cost-efficient screening procedures that can determine whether a patient is predisposed to a disease. While this type of screening provides the obvious benefit of early intervention for diseases such as cancer, Alzheimer's disease, and polycystic kidney disease, it also raises serious questions about how this information may be used by employers and insurance companies.

Vital Data

Tim Beardsley

I f the biotechnology company called Myriad Genetics has its way, thousands of healthy women in the U.S. will hear doubly bad news. First, a close relative—perhaps a sister—will announce that she has breast cancer. Second, the patient's physician thinks this particular cancer has probably been caused by a mutation that the healthy relative has an even chance of also carrying. This patient has been advised to suggest to all her female relatives that they be tested for the mutation. How likely? Hard to say—the mutation has not yet been thoroughly studied—but the likelihood could be as much as 85 percent.

The oncoming tidal wave of genetic data has not yet affected most people. That will change. Workers are now routinely isolating genetic mutations associated with such widespread illnesses as cancer, Alzheimer's disease and some types of cardiovascular disease. Devising tests for mutations in a known gene has become a comparatively straightforward matter. Genzyme, a biotechnology company in Cambridge, Mass., announced in 1995 a diagnostic technology that can simultaneously analyze DNA from 500 patients for the presence of 106 different mutations on seven genes.

When enough is known about the effects of mutations, test results can be a medical boon: they can indicate how likely a person is to develop illnesses and perhaps suggest life-enhancing medical surveillance or therapy. But learning about the effects of mutations requires lengthy study. And genetic data can cause immediate and life-sapping harm. In particular, it can precipitate detrimental psychological changes, and it can open the door to discrimination.

In the past, genetic discrimination has been largely confined to members of families afflicted with rare conditions showing a clear pattern of inheritance. For example, members of families with Huntington's disease, a fatal neurodegenerative disorder that develops in middle age, have long found it difficult or impossible to obtain health insurance. So far a few hundred of people who are at risk of future illness because of their genetic makeup are known to have lost jobs or insurance, Collins states. Most suffered because a family member had been diagnosed with a condition long known to have a genetic basis. But as the number of genetic tests grows, Francis S. Collins predicts, "we are going to see it happen on a larger scale, since we're all at risk for something."

What's Your Genotype?

Testing for cancer-associated mutations, for example, is at present carried out only in research studies at large medical centers, because interpreting the results is fraught with uncertainty. But some grim facts are clear. In families with hereditary breast cancer—which accounts for less than 10 percent of all cases—mutations in the *BRCA1* gene confer an 85 percent lifetime risk of the disease, as well as a 45 percent chance of ovarian cancer. Some women in such families who have learned that they carry a mutated *BRCA1* have elected to undergo a prophylactic mastectomy and ophorectomy (removal of the ovaries)—a procedure that may reduce but does not eliminate the risk of cancer.

More uncertainty arises, however, with women who have a mutated BRCA1 gene but do not have a family history of breast cancer. For them, the danger is uncertain, but it may be smaller. Nor is it known whether the danger is different for members of different ethnic groups (although Ashkenazi Jews are more likely than others to carry one specific mutation in *BRCA1*, and a gene that can cause neurobromatosis has more severe effects in whites than in blacks). These and other uncertainties pose an agonizing treatment dilemma. The choice between radical surgery and intensive surveillance—in the form of frequent mammograms—might be crucial. The identification of a second breast cancer gene, *BRCA2*, complicates matters even more.

Although further advances in understanding the genome might conceivably one day eliminate such dilemmas, most scientists do not expect them to evaporate in the foreseeable future. For a woman who has already been diagnosed with breast cancer, the significance of a positive *BRCA1* test for treatment is murky. And even with a negative test result, a woman still faces the same one-in-eight lifetime risk that all women in the U.S. do. These factors have led the American Society of Human Genetics and the National Breast Cancer Coalition, an advocacy group, to urge that for now, testing for *BRCA1* mutations be carried out only in a research setting. "We will fight any sale of this test before there is consensus on how it should be used," says coalition member Mary Jo Ellis Kahn, a breast cancer survivor with a family history of the disease.

Some women who know that they carry a mutated *BRCA1* gene have gone to elaborate lengths to conceal that information from their insurance carriers, says Barbara B. Biesecker of the NCHGR. One fear is that insurers will classify the mutations as a preexisting condition and so refuse to cover treatments related to the condition. That concern is hardly irrational: health insurance companies often decline to cover policies, or offer inflated premiums, to individuals who have a significant family history of cancer. The National Breast Cancer Coalition was rejected several times for health insurance for its Washington, D.C., staff of eight, because the staff includes some breast cancer survivors. To find coverage, it had to join a larger organization.

The trend toward secrecy in genetic testing seems to be catching on. At a meeting in 1995 of the American Society of Human Genetics, Thomas H. Murray of Case Western Reserve University asked his audience whether they knew of patients who had requested testing for a genetic trait anonymously or under a false name. Hands shot up all over the room. In many clinical studies, patients are now formally warned that test results could lead to insurance complications if they get into the patient's medical records. Researchers sometimes obtain special legal documents called "certificates of confidentiality" that prevent the courts from gaining access to data gathered for a study.

Keep a Secret?

Other patients are simply avoiding taking genetic tests, thus forgoing whatever medical benefit they might bring. People with vonHippelLindau (VHL) disease, a rare hereditary condition that can cause brain and kidney tumors, often find it hard to obtain health insurance because of the expensive surgeries they might need. Although no prophylactic therapy can prevent the tumors, people with VHL disease can extend their lives by undergoing regular magnetic resonance imaging scans followed by surgical removal of tumors. According to William C. Dickson, research management chair of the VHL Family Alliance, many parents with the syndrome avoid having their children tested for mutations in the recently discovered gene for VHL because they fear that a genetic diagnosis will make their offspring uninsurable.

Parents with polycystic kidney disease, which may be the most common simply inherited, life-threatening condition, also frequently decide for insurance-related reasons not to subject their children to testing, reports Gregory G. Germino, an investigator at the Johns Hopkins University School of Medicine. Approximately 600,000 Americans have the illness— many unknowingly. A gene causing many cases, PKD1, was identified in 1994 using technologies developed under the genome project. Testing of PKD1 can sometimes improve medical therapy for a child, Germino says.

Such reports have prompted alarm among health officials. (Collins admits to being passionate on the subject.) A working group on the ethical, legal and social implications of the human genome program, together with the National Action Plan on Breast Cancer, a presidential initiative, recommended that insurance providers be prohibited from using genetic information, or an individual's request for testing, as a basis for limiting or denying health insurance.

Currently insurers do not usually ask directly for results of genetic tests; inquiries about the health or cause of death of a person's parents are sufficient to identify many of those at high risk. But insurers may consider genetic data for an individual policy. They don't ask now about genetic testing, but that will change, says Nancy S. Wexler, president of the Hereditary Disease Foundation, who has herself a 50 percent risk of Huntington's disease.

Because patients with genetic diseases are often reluctant to identify themselves, gauging the extent of discrimination is difficult. But new data strengthen earlier anecdotal reports suggesting the phenomenon is widespread. In one of the first extensive surveys, Lisa N. Geller of Harvard Medical School and her co-authors describe how they sent questionnaires to people who, though free of any symptoms, are at risk for acquiring a genetically based illness. Of the 917 who responded, a total of 455 asserted that they had been discriminated against after they revealed a genetic diagnosis.

Follow-up interviews by the researchers provided details of health and life insurers who refused or canceled coverage, adoption agencies that required prospective parents to pass a genetic test (but on one occasion misunderstood the results) and employers who fired or refused to hire on the basis of a treatable genetic condition or the mere possibility of one. Paul R. Billings of the Veteran's Affairs Medical Center in Palo Alto, Calif., one of the study's authors, declares that the public will reject genetic testing out of fear of discrimination. In a separate study by E. Virginia Lapham of Georgetown University and others, 22 percent of a group of 332 people who had a genetic illness in their families reported having been refused health insurance.

Several European countries have taken steps to prevent abuse of genetic data. Basic medical insurance is not a major concern in Europe, because it is guaranteed by governments. Yet France, Belgium and Norway all have laws preventing the use of genetic information by life and medical insurance companies and by most employers. The Netherlands guarantees basic life insurance, and Germany has some protections. In the U.S. several states have enacted legislation that limits discrimination based on genetic data. Employment discrimination is prohibited by federal law, and several bills now before Congress would discourage or prevent gene-based insurance discrimination nationally. But their prospects are uncertain.

The potential for psychological harm from DNA testing is receiving growing attention. Because a test may have implications for all the members of an extended family, powerful feelings of guilt and sadness can disrupt relationships. Fear of such consequences may explain the unexpectedly low utilization of a test that has been available for some years to identify most carriers of cystic fibrosis—who are not themselves at risk.

Genetic counselors have formed a strong consensus that because of the potential for harm, children should not be tested for mutations predicting diseases that will not develop until adulthood, unless there are possible medical interventions. That principle rules out testing children for Huntington's, because there is no preventive therapy against developing the rocking motions and mental impairment that characterize the illness. Yet parents do seek testing of their children: in one case, to avoid paying for a college education if the youngster was likely to succumb.

The principal U.S. network of testing laboratories is known as Helix. According to a survey by Dorothy C. Wertz and Philip R. Reilly of the Shriver Center for Mental Retardation in Waltham, Mass., 23 percent of the network's labs technically capable of checking for the Huntington's mutation have done so in children younger than 12 years. More than 40 percent of Helix laboratories had performed tests for patients directly, with no physician involved. Yet the public can easily misunderstand the meaning of genetic diagnoses, Wertz notes. Moreover, she says, many physicians are not well enough informed to be giving genetic advice.

On the credit side of the ledger, it is clear that some patients in families afflicted with hereditary colon cancer, and possibly breast cancer, too, have already made wise medical choices as a result of discoveries facilitated by the genome program. Some with a mutation predisposing them to colon cancer, for example, have had their colons removed as soon as threatening changes started to occur—a procedure that probably saved their lives. Eagerly awaited novel therapies, though, are further in the future.

Collins notes that only six years after a team that he co-led found the gene associated with cystic fibrosis in 1989, drugs developed to counter effects of the mutated gene were already being tested in patients. How soon a definitive treatment will emerge, however, is anyone's guess.

The treatment prospect that has most gripped the public imagination is gene therapy—an approach that would be better described as gene transplantation. But attempts to treat familial hypercholesterolemia, cystic fibrosis and Duchenne's muscular dystrophy have each resulted in failures over the past year, apparently because patients' cells did not take up enough of the transplanted genes. The earlier treatment of adenosine deaminase deficiency by W. French Anderson showed at best a modest effect. In December 1995, a NIH review concluded that clinical efficacy has not been definitively demonstrated at this time in any gene therapy protocol.

Following the Money

In spite of these problems, the burgeoning genetic revolution is already causing seismic reverberations in the business world. Pharmaceutical companies have staked hundreds of millions of dollars on efforts to discover genes connected to disease, because they could show the way to molecules that might then be good targets for drugs or diagnostic reagents.

The prospect of commercial exploitation of the genome is motivating protests in some quarters. Most of the political flack is being taken by an initiative known as the Human Genome Diversity Project. The diversity project—which is not formally linked to the genome project—aims to study the variations in genetic sequences among different peoples of the world.

Supporters of the diversity project, which was conceived by Luigi Luca Cavall-Sforza of Stanford University, note that the sequence generated by the genome project proper will be derived mainly from DNA donors of European and North American origin. Studying the 0.1 percent variation among people around the world might yield valuable information about adaptation, Cavalli-Sforza reasons.

Henry T. Greely, a Stanford law professor who is a co-organizer of the genome diversity project, recognizes that data about genetic variation could invite racists to concoct arbitrary rationales justifying discrimination. But he says the project will accept its responsibility to fight such abuses, and he notes that the available data point to how superficial the racial differences are: most of the 0.1 percent of variation in humans occurs among members of the same race, rather than among races.

Still, the project is struggling politically. One of the thorns in its side is the Rural Advancement Foundation International (RAFI), a small campaigning organization based in Ottawa that opposes patents on living things. Jean Christie of the foundation says her group fights the diversity project because it will produce cell lines that can be patented by gene-bagging companies from rich countries.

Greely insists that the project's protocols eliminate the possibility that samples will be used for profit without the consent of donors. Notwithstanding that assurance, a committee of the United Nations Educational, Scientific and Cultural Organization (UNESCO) has criticized the project's lack of contact with indigenous groups during its planning phase. Concerns about exploitation have also been fueled by a controversy over a patent granted to the NIH on a cell line derived from a Papua, New Guinean man. Such worries may explain why the genome diversity project has so far failed to obtain large-scale funding.

How the genome patent race will play out is still unclear. Collins says that a scramble to patent every sequenced gene would be destabilizing, as it would imperil cooperation among investigators. Traditionally, scholars have been free to carry out research unhampered by patents. That freedom cannot be taken for granted, states Rebecca Eisenberg, a patent expert at the University of Michigan. As more scholars forge commercial ties, proprietary interests in the genome may be more vigorously enforced.

Before a gene can be patented, the inventor has to know something about its function, in order to meet the legal requirement of utility. Industry, however, controls most of the research muscle that can efficiently discover useful properties. So commercialization seems to be an inevitable consequence of the genome's scientific exploration, as it is for other explorations. Although the U.S. Patent Office held hearings to examine questions raised by gene patenting, turning back the clock to disallow such patents, as some critics urge, seems unlikely. And so long as corporate dollars do not stiffle collaboration, people in rich countries will probably benefit from the feeding frenzy.

How much the rest of the world will gain, though, is a valid question. Some gene therapies now being evaluated would be tailored to individual patients. But James V. Neel, a pioneer in human genetics at the University of Michigan, warned researchers recently that individual therapies will be too expensive for widespread use. Humble interventions such as improved diet and exercise may ameliorate adult-onset diabetes more cost-effectively than genetic medicine, he points out. Neel urged geneticists to pay attention to the deteriorating environment many humans inhabit as well as to their DNA.

Still, the gene race is on. Better medicines will be found; some people will make fortunes, and some will probably suffer harm. But it is a safe bet that although all humans share DNA, not all of them will share in its bounty. The World Health Organization reports that 12.2 million children under the age of five died in the developing world in 1993. More than 95 percent of those deaths could have been avoided, according to the agency, if those children had access to nutrition and medical care that are already standard practice in countries that can afford them. For many of the world's unfortunates, genetic medicine may always be a distant dream.

One downside of genetic screening for disease is the possibility of genetic discrimination. An insurance company that learns of a person's predisposition for a disease may decline to insure that person. Or an employer may be reluctant to pay premiums for employees who have an elevated risk for a disease that is expensive to treat. What is a person to do?

n all new drugs must ultimat
ted in mammals—and that ofte
e. Mice are very close to h
ns of their genome: more tha
cent of the mouse proteins i
far s   things to kno
teins. Ten laboratories acr
called the Mouse Genome Seq
work, collectively received
n the National Institutes o
t year to lead an effort to

Genetics Discrimination: Pink Slip in Your Genes

Diane Martindale

In April 1999 Terri Seargent went to her doctor with slight breathing difficulties. A simple genetic test confirmed her worst nightmare: she had alpha-1 deficiency, meaning that she might one day succumb to the same respiratory disease that killed her brother. The test probably saved Seargent's life—the condition is treatable if detected early—but when her employer learned of her costly condition, she was fired and lost her health insurance.

Seargent's case could have been a shining success story for genetic science. Instead it exemplifies what many feared would happen: genetic discrimination. A recent survey of more than 1,500 genetic counselors and physicians conducted by social scientist Dorothy C. Wertz at the University of Massachusetts Medical Center found that 785 patients reported having lost their jobs or insurance because of their genes. "There is more discrimination than I uncovered in my survey," says Wertz, who presented her findings at the American Public Health Association meeting in Boston in November, 2000. Wertz's results buttress an earlier Georgetown University study in which 13 percent of patients surveyed said they had been denied or let go from a job because of a genetic condition.

Such worries have already deterred many people from having beneficial predictive tests, says Barbara Fuller, a senior policy adviser at the National Human Genome Research Institute (NHGRI), where geneticists unveiled the human blueprint. For example, one third of women contacted for possible inclusion in a recent breast cancer study refused to participate because they feared losing their insurance or jobs if a genetic defect was discovered. A 1998 study by the National Center for Genome Resources found that 63 percent of people would not take genetic tests if employers could access the results and that 85 percent believe employers should be barred from accessing genetic information. So far genetic testing has not had much effect on health insurance. Richard Coorsh, a spokesperson for the Health Insurance Association of America, notes that health insurers are not interested in genetic tests, for two reasons. First, they already ask for a person's family history—for many conditions, a less accurate form of genetic testing. Second, genetic tests cannot—except for a few rare conditions such as Huntington's disease—predict if someone with a disease gene will definitely get sick.

Public health scientist Mark Hall of Wake Forest University interviewed insurers and used fictitious scenarios to test the market directly. He found that a presymptomatic person with a genetic predisposition to a serious condition faces little or no difficulty in obtaining health insurance. "It's a non-issue in the insurance market," he concludes. Moreover, there is some legislation against it. Four years ago the federal government passed the Health Insurance Portability and Accountability Act (HIPAA) to prevent group insurers from denying coverage based on genetic results. A patchwork of state laws also prohibit insurers from doing so.

Genetic privacy for employees, however, has been another matter. Federal workers are protected to some degree; in February 2000, President Bill Clinton signed an executive order forbidding the use of genetic testing in the hiring of federal employees. But this guarantee doesn't extend to the private sector. Currently an employer can ask for, and discriminate on the basis of, medical information, including genetic test results, between the time an offer is made and when the employee begins work. A 1999 survey by the American Management Association found that 30 percent of large and midsize companies sought some form of genetic information about their employees, and 7 percent used that information in awarding promotions and hiring. As the cost of DNA testing goes down, the number of businesses testing their workers is expected to skyrocket.

Concerned scientists, including Francis S. Collins, director of the NHGRI and the driving force behind the Human Genome Project, have called on the Senate to pass laws that ban employers from using DNA testing to blacklist job applicants suspected of having "flawed" genes. Despite their efforts, more than 100 federal and state congressional bills addressing the issue have been repeatedly shelved in the past two years. "There is no federal law on the books to protect [private-sector] employees, because members of Congress have their heads in the sand," contends Joanne Hustead, a policy director at the National Partnership for Women and Families, a nonprofit group urging support of federal legislation. "Your video rental records are more protected," she claims.

Wertz also believes that more laws are simply Band-Aids on the problem: "We need a public health system to fix this one." And she may be right. In nations such as Canada and the U.K., where a national health service is in place, the thorny issue of genetic discrimination is not much of a concern.

While policymakers play catch-up with genetic science, Seargent and others are hoping that the Equal Employment Opportunity Commission (EEOC) will help. The EEOC considers discrimination based on genetic traits to be illegal under the Americans with Disabilities Act of 1990, which safeguards the disabled from employment-based discrimination. The commission has made Seargent its poster child and is taking her story to court as a test case on genetic discrimination.

Seargent, who now works at home for Alpha Net, a Web-based support group for people with alpha-1 deficiency, doubts she'll be victorious, because all but 4.3 percent of ADA cases are won by the employer. She does not regret, however, having taken the genetic test. "In the end," she says, "my life is more important than a job." Ideally, it would be better not to have to choose.

The aim of the Human Genome Project is to seek information about ourselves that is locked up within our DNA molecules. This information assembles our bodies from a single fertilized egg cell. It tells what diseases we may be susceptible to and what medicines might be able to cure them; it even tells us how we might design and create the medicines. It tells who our parents were and who our ancestors might have been, all the way back to the origin of multicellular life. It tells of our relationships with all of earth's other life forms, from bacteria to oak trees to chimpanzees. And it tells what makes us human, if we have the wit to read it correctly.

Bioinformatics is the discipline of deciphering the sequenced genome, that is, how to harvest or mine all the data once the sequence is in hand. The following essay covers various aspects of bioinformatics, including some of the business and political considerations involved in conducting a project of this magnitude. Eventually, it might even become possible to simulate an entire genome with a supercomputer, so that we could create a "virtual life form" in cyberspace and watch it live.

all new drugs must ultimat
ted in mammals—and that ofte
e. Mice are very close to h
ns of their genome: more tha
cent of the mouse proteins i
ar s   to kno
eins. Ten laboratories acr
called the Mouse Genome Seq
work, collectively received
the National Institiutes o

The Bioinformatics Gold Rush

Ken Howard

P lastics." When a family friend whispered this word to Dustin Hoffman's character in the 1967 film *The Graduate*, he was advocating not just a novel career choice but an entirely different way of life. If that movie were made today, in the age of the deciphering of the human genome, the magic word might well be "bioinformatics."

Corporate and government-led scientists have already compiled the three gigabytes of paired A's, C's, T's and G's that spell out the human genetic code—a quantity of information that could fill more than 2,000 standard computer diskettes. But that is just the initial trickle of the flood of information to be tapped from the human genome. Researchers are generating gigantic databases containing the details of when and in which tissues of the body various genes are turned on, the shapes of the proteins the genes encode, how the proteins interact with one another and the role those interactions play in disease. Add to the mix the data pouring in about the genomes of so-called model organisms such as fruit flies and mice, and you have what Gene Mayers, Jr., vice president of informatics research at Celera Genomics in Rockville, Md., calls "a tsunami of information." The new discipline of bioinformatics—a marriage between computer science and biology—seeks to make sense of it all. In so doing, it is destined to change the face of biomedicine.

"For the next two to three years, the amount of information will be phenomenal, and everyone will be overwhelmed by it," Myers predicts. "The race and competition will be who can mine it best. There will be such a wealth of riches."

A whole host of companies are vying for their share of the gold. Jason Reed of the investment banking firm Oscar Gruss & Son in New York City estimates that bioinformatics could be a $2-billion business by 2005. He has complied information on more than 50 private and publicly traded companies that offer bioinformatics products and services. These companies plug into the effort at various points: collecting and storing data, searching databases, and interpreting the data. Most sell access to their information to pharmaceutical and biotechnology companies for a hefty subscription price that can run into the millions of dollars.

The reason drug companies are so willing to line up and pay for such services—or to develop their own expensive resources in-house—is that bioniformatics offers the prospect of finding better drug targets earlier in the drug development process. This efficiency could trim the number of potential therapeutics moving through a company's clinical testing pipeline, significantly decreasing overall costs. It could also create extra profits for drug companies by whittling the time it takes to research and develop a drug, thus lengthening the time a drug is on the market before its patent expires.

"Assume I'm a pharmaceutical company and somebody can get [my] drug to the market one year sooner," explains Stelios Papadopoulos, managing director of health care at the New York investment banking firm SG Cowen. "It could mean you could grab maybe $500 million in sales you would not have recovered."

Before any financial windfalls can occur, however, bioinformatics companies must contend with the current plethora of genomic data while constantly refining their technology, research approaches and business models. They must also focus on the real challenge and opportunity—finding out how all the shards of information relate to one another and making sense of the big picture.

"Methods have evolved to the point that you can generate lots of information," comments Michael R. Fannon, vice president and chief information officer of Human Genome Sciences, also in Rockville. "But we don't know how important that information is."

Divining that importance is the job of bioinformatics. The field got its start in the early 1980s with a database called Gen-Bank, which was originated by the U.S. Department of Energy to hold the short stretches of DNA sequence that scientists were just beginning to obtain from a range of organisms. In the early days of GenBank a roomful of technicians sat at keyboards consisting of only the four letters A, C, T and G, tediously entering the DNA-sequence information published in academic journals. As the years went on, new protocols enabled researchers to dial up GenBank and dump in their sequence data directly, and the administration of GenBank was transferred to the National Institutes of Health's National Center for Biotechnology Information (NCBI). After the advent of the World Wide Web, researchers could access the data in GenBank for free from around the globe.

Once the Human Genome Project (HGP) officially got off the ground in 1990, the volume of DNA-sequence data in GenBank began to grow exponentially. With the introduction in the 1990s of high-throughput sequencing—an approach using robotics, automated DNA-sequencing machines and computers—additions to GenBank skyrocketed. GenBank held the sequence data on more than seven billion units of DNA by July 2000.

Around the time the HGP was taking off, private companies started parallel sequencing projects and established huge proprietary databases of their own. Today companies such as Incyte Genomics in Palo Alto, Calif., can determine the sequence of approximately 20 million DNA base pairs in just one day. And Celera Genomics says that it has 50 terabytes of data storage. That's equivalent to roughly 80,000 compact discs, which in their plastic cases would take up almost half a mile of shelf space.

But GenBank and its corporate cousins are only part of the bioinformatics picture. Other public and private databases contain information on gene expression (when and where genes are turned on), tiny genetic differences among individuals called single-nucleotide polymorphisms (SNPs), the structures of various proteins, and maps of how proteins interact.

Mixing and Matching

One of the most basic operations in bioinformatics involves searching for similarities, or homologies, between a newly sequenced piece of DNA and previously sequenced DNA segments from various organisms. Finding near-matches allows researchers to predict the type of protein the new sequence encodes. This not only yields leads for drug targets early in drug development but also weeds out many targets that would have turned out to be dead ends.

A popular set of software programs for comparing DNA sequences is BLAST (for Basic Local Alignment Search Tool), which first emerged in 1990. BLAST is part of a suite of DNA- and protein-sequence search tools accessible in various customized versions from many database providers or directly through NCBI. NCBI also offers Entrez, a so-called metasearch tool that covers most of NCBI's databases, including those housing three dimensional protein structures, the complete genomes of organisms such as yeast, and references to scientific journals that back up the database entries.

An early example of the utility of bioinformatics is cathepsin K, an enzyme that might turn out to be an important target for treating osteoporosis, a crippling disease caused by the breakdown of bone. In 1993 researchers at SmithKline Beecham, based in Philadelphia, asked scientists at Human Genome Sciences to help them analyze some genetic material they had isolated from the osteoclast cells of people with bone tumors. (Osteoclasts are cells that break down bone in the normal course of bone replenishment; they are thought to be overactive in individuals with osteoporosis.)

Human Genome Sciences scientists sequenced the sample and conducted database homology searches to look for matches that would give them a clue to the proteins that the sample's gene sequences encoded. Once they found near-matches for the sequences, they carried out further analyses and discovered that one sequence in particular was overexpressed by the osteoclast cells and that it matched those of a previously identified class of molecules: cathepsins.

For SmithKline Beecham, that exercise in bioinformatics yielded in just weeks a promising drug target that standard laboratory experiments could not have found without years and a pinch of luck. Company researchers are now trying to find a potential drug that blocks the cathepsin K target. Searches for compounds that bind to and have the desired effect on drug targets still take place mainly in a biochemist's traditional "wet" lab, where evaluations for activity, toxicity and absorption can take years. But with new bioinformatics tools and growing amounts of data on protein structures and biomolecular pathways, some researchers say, this aspect of drug development will also shift to computers, in what they term "in silico" biology.

It all adds up to good days ahead for bioinformatics, which

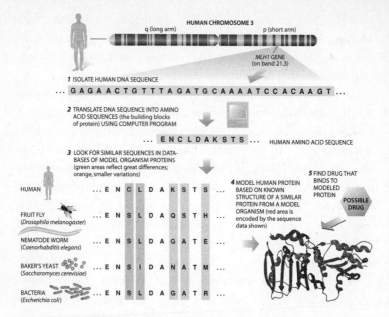

Using Bioinformatics to Find Drug Targets

By looking for genes in model organisms that are similar to a given human gene, researchers can learn about the protein the human gene encodes and search for drugs to block it. The *MLHI* gene, which is associated with colon cancer in humans, is used in this example.

many assert holds the real promise of genomics. "Genomics without bioinformatics will not have much of a payoff," states Roland Somogyi, former director of neurobiology at Incyte Genomics who is now at Molecular Mining in Kingston, Ontario.

Michael N. Liebman, head of computational biology at Roche Bioscience in Palo Alto, agrees. "Genomics is not the paradigm shift; it's understanding how to use it that is the paradigm shift," he asserts. "In bioinformatics, we're at the beginning of the revolution."

The revolution involves many different players, each with a different strategy. Some bioinformatics companies cater to large users, aiming their products and services at genomics, biotechnology and pharmaceutical companies by creating custom software and offering consulting services. Lion Bioscience, based in Heidelberg, Germany, has been particularly successful at selling "enterprise-wide" bioinformatics tools and services. Its $100-million agreement with Bayer to build and manage a bioinformatics capability across all of Bayer's divisions was at that time the industry's largest such deal.

Other firms target small or academic users. Web businesses such as Oakland, Calif.-based Double Twist and eBioinformatics, which is headquartered in Pleasanton, Calif., offer one-stop Internet shopping. These on-line portals allow users to access various types of databases and use software to manipulate the data.

In May 2000, DoubleTwist scientists announced they had used their technology to determine that the number of genes in the human genome is roughly 105,000, although they said the final count would probably come in at 100,000. For those who would rather have the software behind their own security firewalls, Informax in Rockville, Oxford Molecular Group in England, and others sell shrink-wrapped products.

Making Connections

Large pharmaceutical companies—"big pharma"—have also sought to leverage their genomics efforts with in-house bioinformatics investments. Many have established entire departments to integrate and service computer software and facilitate database access across multiple departments, including new product development, formulation, toxicology and clinical testing. The old model of drug development often compartmentalized these functions, ghettoizing data that might have been useful to other researchers. Bioinformatics allows researchers across a company to see the same thing while still manipulating the data individually.

In addition to making drug discovery more efficient, in-house bioinformatics can also save drug companies money in software support. Glaxo Wellcome in Research Triangle Park, N.C., is replacing individual packages used by various investigators and departments to access and manipulate databases with a single software platform. Robin M. DeMent, U.S. director of bioinformatics at Glaxo Wellcome, estimates that this will save approximately $800,000 in staffing support over a three-to five-year period.

To integrate bioinformatics throughout their companies, pharmaceutical giants also forge strategic alliances, enter into licensing agreements and acquire smaller biotechnology companies. Using partners and vendors not only allows big pharma to fill in the gaps in its bioinformatics capabilities but also gives it the mobility to adapt new technologies as they come onto the market rather than constantly overhauling its own systems. "If a pharmaceutical company had a large enough research budget, they could do it all themselves," Somogyi says. "But it's also a question of culture. The field benefits as a whole by providing different businesses with different roles with room to overlap."

Occupying some of that overlap—in resources, products and market capitalization—are companies such as Human Genome Sciences, Celera and Incyte. They straddle the terrain between big pharma and the data integration and mining offered by specialist companies. They have also quickly seized on the degree of automation that bioinformatics has brought to biology.

But with all this variety comes the potential for miscommunication. Getting various databases to talk to one another—what is called interoperability—is becoming more and more key as users flit among them to fulfill their needs. An obvious solution would be annotation—tagging data with names that are cross-referenced across databases and naming systems. This has worked to a degree. "We've been successful in bringing databases together by annotation: database A to database

B, B to C, C to D," explains Liebman of Roche Bioscience. "But annotation in A may change, and by the time you get down to D the references may not have changed, especially with a constant stream of new data." He points out that this problem becomes more acute as the understanding of the biology and the ability to conduct computational analysis becomes more sophisticated. "We're just starting to identify complexities in these queries, and how we store data becomes critical in the types of questions we can ask," he states.

Systematic improvements will help, but progress—and ultimately profit—still relies on the ingenuity of the end user, according to David J. Lipman, director of NCBI. "It's about brainware," he says, "not hardware or software."

> *One of the hardest parts about sharing genetic information is finding a common language. Researchers from biotechnological and pharmaceutical companies are attempting to devise computer standards that would allow genetic scientists to share data. An immediate benefit of creating such a standard would be the increased speed in development of gene-specific drugs. But which language should be used?*

pharmaceutical company

HGS; Celera; Incyte } Bioinformatics

[L 3 4 5 6 7 etc

data integration specialist

Hooking Up Biologists

Carol Ezzell

I magine that your co-worker in the next cubicle has some information you need for a report that's due soon. She e-mails it to you, but the data are from a spreadsheet program, and all you have is a word processor, so there's no possibility of your cutting and pasting it into your document. Instead you have to print it out and type it in all over again. That's roughly the situation facing biologists these days. Although databases of biological information abound—especially in this post-genome-sequencing era—many researchers are like sailors thirsting to death surrounded by an ocean: what they need is all around them, but it's not in a form they can readily use.

To solve the problem, various groups made up of academic scientists and researchers from biotechnology and pharmaceutical companies are coming together to try to devise computer standards for bioinformatics so that biologists can more easily share data and make the most of the glut of information resulting from the Human Genome Project. Their goal is to enable an investigator not only to float seamlessly between the enormous databases of DNA sequences and those of the three-dimensional protein structures encoded by that DNA. They also want a scientist to be able to search the databases more efficiently so that, to use an automobile metaphor, if someone typed in "Camaro," the results would include other cars as well because the system would be smart enough to know that a Camaro is another kind of car.

The immediate payoff is expected to be the faster development of new drugs. "Pharmaceutical research is the only industry I know of with declining productivity," says Tim Clark, vice president of informatics for Millennium Pharmaceuticals in Cambridge, Mass. "The R&D effort is at a primitive craft scale, like cottage weavers, although standardization is one of the first problems that got tackled in the Industrial Revolution, with the invention of interchangeable parts."

The issue is what standards to use. In a situation reminiscent of the computer industry in the 1970s, everyone advocates standards, as long as they are his or her own. Formal groups have sprung up worldwide with names like the BioPathways Consortium, the Life Sciences Research Domain Task Force of the Object Management Group, and the Bio-Ontologies Consortium—and each has a different idea of how things should be done. Eric Neumann, a member of both the Bio-Ontologies and BioPathways consortia, is a neuroscientist who is now vice president for life science informatics at the consulting firm 3rd Millennium in Cambridge, Mass. (no relation to Millennium Pharmaceuticals). He says Extensible Markup Language (XML) is shaping up to be the standard computer language for bioinformatics. XML is the successor to Hypertext Markup Language (HTML), the current driver of the World Wide Web.

One of XML's advantages is that it contains tags that identify each kind of information according to its type: "Camaro," for example, would be tagged as a car. Neumann proposes that XML-based languages will "emphasize the Web-like nature of biological information," which stretches from DNA to messenger RNA, proteins, protein-protein interactions, biochemical pathways, cellular function and, ultimately, the behavior of a whole organism. Current ways of storing and searching such biological information are centered on single genes, according to Neumann, "but the diseases we want to treat involve more than one gene."

Clark says the main problems facing bioinformatics that make standard development necessary are the sheer volume of data, the need for advanced pattern recognition (such as within DNA sequences and protein structural domains), the ability to process signals to eliminate "noise" from data, and something called combinatorial optimization, or finding the best path through a maze of molecular interactions. "You can't build all of it yourself," he contends.

Neumann thinks combinatorial optimization could be the highest hurdle. "Pathways are a lot more complex than [DNA] sequences," he states. "If we don't come up with something, it's going to be a real mess."

In the following interview, Stuart Kauffman, chief scientific officer and co-founder of Cistem Molecular and leading entrepreneur in the developing field of bioinformatics, discusses how computers may be used to determine the circuitry and logic of genes and cells.

all new drugs must ultimat
ed in mammals—and that ofte
.   Mice are very close to h
ns of their genome: more tha
cent of the mouse proteins i
far similar enough to kno
teins.   Ten laboratories acr
called the Mouse Genome Seq
work, collectively received
the National Institiutes o

Interview with Stuart Kauffman

Kristin Leutwyler

Stuart Kauffman wears many hats. He is an entrepreneur who founded Bios Group, a Santa Fe-based software company where he is now chief scientific officer and chairman of the board, and co-founded Cistem Molecular in San Diego. He is an academic, with current posts as an external professor at the Santa Fe Institute and a professor emeritus at the University of Pennsylvania School of Medicine. And he is an author, having written numerous papers and three popular books (*Origins of Order: Self-Organization and Selection in Evolution*, *At Home in the Universe* and *Investigations*). But perhaps most of all, he is a visionary.

Indeed, Kauffman is among the pioneering scientists now mapping the intersection of computational mathematics and molecular biology—a burgeoning field known as bioinformatics. *Scientific American* sat down to discuss this relatively new discipline.

SA: What is the promise of bioinformatics?
The completion of the human genome project itself is a mar-

velous milestone, and it's the starting gate for what will become the outstanding problem for the next 15 to 20 years in biology—the postgenomic era that will require lots of bioinformatics. It has to do with how we understand the integrated behavior of 100,000 genes, turning one another on and off in cells and between cells, plus the cell signaling networks within and between cells. We are confronting for the first time the problem of integrating our knowledge.

SA: How so?

The fertilized egg has somewhere between 80,000 and 100,000 structural genes. I guess we'll know pretty quickly what the actual answer is. We're entitled to think of the, let's say, 100,000 genes in a cell as some kind of parallel processing chemical computer in which genes are continuously turning one another on and off in some vastly complex network of interaction. Cell signaling pathways are linked to genetic regulatory pathways in ways we're just beginning to unscramble. In order to understand this, molecular biologists are going to have to—and they are beginning to—change the way they think about cells. We have been thinking one gene, one protein for a long time, and we've been thinking very simple ideas about regulatory cascades called developmental pathways.

The idea is that when cells are undergoing differentiation, they are following some developmental pathway from a precursor cell to a final differentiated cell. And it's true that there's some sort of pathway being followed, but the relationship between that and the rolling change of gene activities is far from clear. That's the huge problem that's confronting us. So the most enormous bioinformatics project that will be in front of us is unscrambling this regulatory network. And it's not going to merely be bioinformatics; there has to be a marriage between new kinds of mathematical tools to solve this problem. Those tools will in general suggest plausible alternative circuits for bits and pieces of the regulatory network. And then we're going to have to marry that with new kinds of experiments to work out what the circuitry in cells actually is.

SA: *Who is going to be working on this entire rubric? Is it bioinformaticians, or is it mathematicians or biologists?*
All of the above. As biologists become aware of the fact that this is going to be essential, they are beginning to turn to computational and mathematical techniques to begin to look at it. And meanwhile we have in front of us the RNA chip data that's becoming available and proteomics as well. An RNA chip shows the relative abundance of transcripts from a very large number of different genes that have been placed on the chip from a given cell type or a given tissue sample. There are beginning to be very large databases of RNA chips that have expression data for tens of thousands of genes including normal cells, diseased cells, untreated and treated normal and diseased cells. Most of the data is a single moment snapshot. You just sample some tissue and see what it is doing. But we're beginning to get data for developmental pathways.

So you have a precursor cell that's triggered to differentiate by giving it some hormone. And you take staged samples of the thing as it differentiates, and we watch genes wax and wane in their activities during the course of that differentiation.

That poses a problem of what to do with all that data. Right now what people are doing largely is <u>cluster analysis</u>, which is to say they take the chip data from a bunch of different cell types and try to cluster the genes that are expressed to a high level versus those that are expressed to a low level. And in effect that's simply a different way of recognizing cell types, but now at the molecular level. And there's nothing wrong with doing that, but there's nothing functional about it in the sense that if you're interested in finding out that gene A turns on gene B, and gene B—when it's on—turns off gene C. The way people are going about analyzing it doesn't lead in the direction of answering those questions.

SA: *Why is that?*

Because they're just recognizing patterns. Which is useful for diagnostic purposes and treatment purposes, but it's not the way to use the data to find what the regulatory circuits are.

SA: *What could you do once you discover the circuitry?*

First of all, you've just broadened the target range for the drug industry. Suppose a given gene makes an enzyme, then perhaps the enzyme's a nicely drugable target and you can make a molecule that enhances or inhibits the activity of the enzyme. But something you could do instead would be to turn on or off the gene that makes the enzyme. By finding the circuitry in and around a gene of medical interest, you've just expanded the number of drugable targets, so that you can try to modulate the activity of the genetic network rather than impinging upon the product of the gene.

Also, anything along the lines of diagnostics, if I know patterns of gene activities and regulatory circuitries, I can test to see the difference between a normal cell type and a hepatic cancer cell—a liver cancer cell. That is obviously useful diagnostically and therapeutically.

The biggest and longest-term consequences of all of this is uncovering the genetic regulatory network that controls cell development from the fertilized egg to the adult. That means that in the long run, we're going to be able to control cell differentiation and induce cell death, apoptosis. My dream is the following: 10 or 20 years from now, if you have prostatic cancer, we will be able to give drugs that will induce the cancer cells to differentiate in such a way that they will no longer behave in a malignant fashion, or they'll commit suicide by going into apoptosis. Then we'll also be able to cause tissue regeneration so that if you happen to have lost half of your pancreas, we'll be able to regenerate your pancreas. Or we'll be able to regenerate the beta cell islets in people who have diabetes. After all, if we can clone Dolly the sheep and make a whole sheep from a single cell, and if we now have embryonic stem cells, what we need are the chemical inductive stimuli that can control pathways of differentiation so that we can cause tissue to regenerate much more at will than we can now. I think that's going to be a huge transformation in biomedicine.

SA: What part does bioinformatics play in achieving this?
Most cancer cells are monoclonal; that means they're all derived from some single cell. And most cancer cells when they differentiate are leaky in the sense that they give rise to normal cell types as well as to cancer cells. This is a fact known to oncologists, which is not part of our therapeutic regimen right now. Cancer cells give rise to both normal and cancer cell types when the cancer stem cell, which is maintaining a monoclonal line, is proliferating. What if we could take that cancer cell, give it chemical signals that induce it to differentiate into normal cell types—we would be treating the cancer cell not by killing cells, but by using jujitsu on them and diverting them to be normal cell types. This already works with vitamin A and certain cancer cells, so there's already a precedent for it.

This is just one example of what we'll be able to do as we discover the circuitry and the logic—and therefore the dynamical behavior—of cells. We will know which gene we have to perturb with what, or which sequences of genes we have to perturb in what temporal order, to guide the differentiation of a cancer cell to nonmalignant behavior or to apoptosis, or to guide the regeneration of some tissue. I can imagine the time when we'll be able to regenerate cardiac tissue from surrounding normal tissue instead of having a scar in place, and the scar serves as the focus for getting recurrent electrical activity in the heart, sending up little spirals of electrical activity, which make your heart beat unstably and which makes people postheart attack subject to sudden cardiac death because they go into ventricular fibrillation.

Suppose what we could do instead of getting scar tissue,

suppose we could get those cells to differentiate into perfectly normal myofibrils. Nothing says we can't do that since the muscle cells and fibrotic tissue are cousins of one another developmentally. So you could begin to imagine treating all kinds of diseases by controlling cell differentiation, tissue differentiation and so on. And to do that we're going to have to know what the circuitry is, and we're going to have to know what small molecules or molecules in general can be added to a person that will specifically treat the diseased tissues and not have undue side effects.

SA: How does complexity theory, disorganization/self-organizing systems, come into play? How do computers and algorithms and data from many different places need to be integrated?

There are three ways we will understand genetic regulatory networks, all of which involve computational work, as well as piles of data. One of them has already been pioneered. It is the following: I have a small genetic circuit, for example, bacteriophage lambda—or something like that,—which has 20 or 30 genes in it and one major switch. And I know all of the genes; I know which genes make which products, which bind to which genes; I know the binding constants by which those gene products bind to the gene. And what I do is, I make in effect an engineering model of that specific circuit, sort of like electrical engineering, except it's molecular-biology-chemical engineering to make a specific circuit for that bacterium. One is inclined to do the same thing with the human genome.

Suppose I pick out 10 genes that I know regulate one another. And I try to build a circuit about their behavior. It's a perfectly fine thing, and we should do it. But the downside is the following: those 10 genes have inputs from other genes outside that circuit. So you're taking a little chunk of the circuitry that's embedded in a much larger circuit with thousands of genes in it. You're trying to figure out the behavior of that circuit when you do not know the outside genes it impacted. And that makes that direct approach hard because you never know what the other inputs are. Evidence that it's hard comes from a parallel—looking at neural circuits, at neural ganglia. We've known for years what every neuron is in, say, the lobster gastric ganglia; what all of the synaptic connections are; what the neurotransmitters are; and you have maybe 13 or 20 neurons in the ganglion, and you still can't figure out the behavior of the ganglion. So no mathematician would ever think that understanding a system with 13 variables is going to be an easy thing to do. And we want to do it with 100,000 variables. That scales the problem.

Molecular biologists have thought they're going to be able to work out how 100,000 genes work with one another without having to write down the mathematical equations by which genes govern one another, and then figuring out from that what the behavior is. That's why this is a stunning transition that we're going through, and there's a lot of stumbling around going on. It's because molecular biologists don't know any mathematics, by and large.

The second approach to this problem is one that I've pioneered, and it's got great strengths but great weaknesses. It turns out that if you want to model genes as if they're little lightbulbs, which they're not, but if you want to model them that way, then the appropriate rules that talk about turning genes on and off are called Boolean functions. If you have K inputs to a gene, there's only a finite number of different Boolean functions.

So what I started doing years ago was wondering if you just made genetic networks with the connections at random with the logic that each gene follows is assigned at random, would there be a class of networks that just behaved in ways that looked like real biologic networks? Are there classes of networks where all I do is tell you all I know about the number of inputs, and some biases on the rules by which genes regulate one another—and it turns out that a whole class of networks, regardless of the details, behaves with the kind of order that you see in normal development?

SA: How would you identify the different networks as one class or another?

There are two classes of networks: one class behaves in an ordered regime and the other class behaves in a chaotic regime, and then there is a phase transition, dubbed "the edge of chaos," between the two regimes. The ordered regime shows lots and lots of parallels to the behavior of real genetic systems undergoing real development. To give an example, if I make a network with two inputs per gene and that's all I tell you, and I make a huge network, with 50,000 or 100,000 genes, and everybody's got two inputs, but it's a scrambled mess of a network in terms of the connections—it's a scrambled spaghetti network, and the logic assigned to every gene is assigned at random, so the logic is completely scrambled—that whole system nevertheless behaves with extraordinary order. And the order is deeply parallel to the behavior of real cell types.

Even with 10 inputs per gene, networks pass from the chaotic regime into the ordered regime if you bias the rules with canalizing functions. The data is very good that genes are regulated by canalizing functions. There is one caveat. It could be that among the known genes that are published, it's predominantly the case that they are governed by canalizing functions because such genes have phenotypic effects that are easy to find, and there's lot of things that are noncanalizing functions, but you just can't find them easily, genetically. So one of the things that we'll have to do is take random genes out of the human genome or the yeast genome or the fly genome and see what kind of control rules govern them.

SA: What would be the implication if it did turn out that most were governed by canalizing functions?

What we've done is made large networks of genes, modeled genes mathematically, in which we've biased the control rules to ask whether or not such networks are in the ordered or chaotic regime, and they are measurably in the ordered regime. The implication is that natural selection has tuned the fractions of genes governed by canalizing functions such that cells are in the ordered regime.

The way you do this is you make an ensemble of all possible networks with the known biases, number of input per genes and biases on the rules. And you sample thousands of networks drawn at random from that ensemble, and the conclusions that you draw are conclusions for typical members of the ensemble. This is a very unusual pattern of reasoning in biology. It's precisely the pattern of reasoning that happens in statistics with things like spin glasses, which are disordered magnetic materials. The preeminent place it has been used is in statistical physics. The weakness of this ensemble approach, this ensemble of networks, is that you can never deduce from it that gene A regulates gene F because you're making statistical models. The strength is that you can deduce lots of things about genetic regulatory nets that you can't get to if you make little circuits and try to do the electrical engineering approach.

In the simplest case in these model networks, there's a little central clock that ticks; that's not true for real cells, but it will do for right now. Every gene looks at the state of its inputs and it does the right thing. So let me define the state of the network as the current on and off values of all 100,000 genes. So how many states are there? Well, there are two possibilities for gene 1 and two possibilities for gene 2 and so on, so there are $2^{100,000}$, which is $10^{30,000}$, so we're talking about a system in the human genome, even if we treat genes as idealized as on or off—which is false because they show graded levels of activity—it's got $10^{30,000}$ possible states. It is mind-boggling because the number of particles in the known universe is 10^{80}.

Here's what happens in the ordered regime. At any moment in time, the system is in a state, and there's a little clock; when the clock ticks, all the genes look at the states of their inputs and they do the right thing, so the whole system goes from some state to some state, and then it goes from state to state along a trajectory. There's a finite number of states; it's large, but it's finite. So eventually the system has to hit a state it's been in before, and then it's a deterministic system; it'll do the same thing. So it will now go around a cycle of states. So the generic behavior is a transient that flows into a cycle.

The cycle is called a state cycle or an attractor. What I did 30 years ago was ask, "What's a cell type?" And I guessed that cell types were attractors, because otherwise we'd have $10^{30,000}$ different cell types, and we have something like 260. So here's what happens in the ordered regime. The number of states on the state cycle tends to be about the square root of the number of genes. The square root of 100,000 is around 318. So this system with $10^{30,000}$ states settles down to a little cycle with 300 states on it. That's enormous order. The system had squeezed itself down to a tiny black hole in its states phase.

If you're on one of these attractor state cycles and you perturb the activity of a single gene, like if a hormone comes in, most of the time you come back to the same attractor, so you have homeostasis. Sometimes, however, you leave one state cycle and you jump onto a transient that goes onto another state cycle, so that's differentiation.

All the things I'm telling you are testable things about the human genome. And they are predictions about the integrated behavior of the whole genome, and there's no way of getting to that right now without using the ensemble approach. They're very powerful predictions. That's the strength of it. The weakness is it doesn't tell you that gene A regulates gene F, which of course is exactly one of the things that we want to know.

SA: *What does it buy you?*
There's all sorts of questions you can answer using the ensemble approach that we will not be able to do until we have a complete theory of the genome.

SA: *What then is the next step to take? Take the data from the genome and plug it into these models?*
Yes, in the sense that you can do experiments to test all of the predictions of these kinds of ensemble models. Nothing prevents me from cloning in a controllable promoter upstream from 100 different randomly chosen genes in different cell lines, perturbing the activity of the adjacent gene, and using Affymetrix chips to look at the avalanches of changes of gene activity. All of that is open to be tested.

We should be able to predict not only that it happens but the statistical distribution of how often when you do it cell type A goes back to being cell type A and how often cell type A becomes cell type B. Everything here is testable. In the actual testing of it for real cells, we'll begin to discover which perturbation to which gene actually causes which pathway of differentiation to happen. You can use molecular diversity or combinatorial chemistry to make the molecules with which you do the perturbation of cells and then test the hypotheses we've talked about.

The aim is to try and find means to either change the abundance of a given gene transcript to treat a disease or to cause differentiation. I have more than one friend who either had or has cancer and our methods for treating cancer are a blunderbuss, really idiotic, even though they are much more sophisticated than they used to be. We're just killing dividing cells. What if we could get to where we could direct cells to differentiate? It's huge in its practical importance if we could make that happen.

SA: *Is bioinformatics the tool to integrate the computational work and the wet work?*

Bioinformatics has to be expanded to include experimental design. What we're going to get out of each of these pieces of bioinformatics is hypotheses that now need to be tested. And it helps you pick out what hypothesis to go test. And the reason is we don't know all 100,000 genes and the entire circuitry. Even if we knew the entire circuitry, as we do for ganglia in the lobster gut—people having been working for 30 years to understand how the lobster gut ganglia work, even knowing all the anatomical connections. So it isn't going to be easy.

I think the greatest intellectual growth area will come with the inverse problem. The point is the following: I show you the Affymetrix chips of differing patterns of gene expression and you tell me from that data what gene actually regulates what gene by what logic. That's the inverse problem. I show you the behavior of the system and you deduce the logic and the connections among the genes.

SA: *Do you see being able to do in silico work for the entire human body or various circuits anytime soon, or ever?*

Yes, I do. I think our timescale is 10 to 15 years to develop good models of the circuitry in cells because so much of the circuitry is unknown. But before I can make the model and explore the dynamical behavior of the model, I can either use the ensemble approach, which I've used, or I actually have to know what the circuitry is. There are three approaches for discovering the circuitry. One is purely experimental, which is what molecular biologists have been doing for a long time.

SA: *How accurate is it? How testable is it?*

It works for small networks, for synchronous lightbulb networks. Real cells aren't lightbulbs; they're graded levels of activity and they're not synchronous, so it's a much harder problem to try and do this for real cells for a variety of reasons.

First of all, when you take a tissue sample, you don't have a single cell type, you typically have several cell types in it. A lot of the data that's around has to do with tissue samples because it's from biopsy data. Second, most of it is single-moment snapshots rather than a sequence of gene activities along a developmental pathway. That's harder data to get.

It's beginning to become available. It is from the state transition that we can learn an awful lot of the data about which genes regulate which genes. There are other potentially powerful techniques that amount to looking at correlated fluctuations in patterns of gene activity and trying to work out from those which genes regulate which genes by what rules. The bottom line of such inverse problem efforts is that the algorithms are going to come up with alternative possible circuits that would explain the data. And that then is going to guide you to ask what's the right next experiment to do to reduce your uncertainty about what the right circuit is. So the inverse problem is going to play into the development of experimental design.

SA: You start with microarray experimentation?
You go in saying, "We think gene A regulates gene F." And now you say, "If that's true, if I perturb the activity of gene A, I should see a change in the activity of gene F; specifically, if I turn gene A on, I should turn gene F on." So now you go back and find a cis site and a trans factor, or a small molecule which, when added to the cell, will turn gene A on, and then you use an Affymetrix chip or some analogue of that to say, "Did I just turn on gene F?"

SA: What is the practical application?
We've just expanded the set of drugable targets from the enzyme itself to the circuitry that controls the activity of the gene that makes the enzyme.

SA: What is bioinformatics' role in this entire enterprise?
Let's take the inverse problem. The amount of data that one is

going to need from real fluctuating patterns of gene expression and the attempt to deduce from that which gene regulates which gene by what logic—that's going to require enormous computational power. I thought the problem was going to be what's called NP-hard, namely exponentially hard, in the number of genes. I now think it's not; I think it's polynomially hard, which means it's solvable or it's much more likely to be solvable in the number of genes.

The real reason is the following: Suppose that any given gene has a maximum of somewhere between one and ten inputs. Those inputs, if you think of them as just on or off, can be in 2^{10} states; they can all be on or can all be off or any other combination. Well, 2^{10} is 1,000. That's a pretty big number. But it's small compared to $10^{30,000}$. Since most genes are regulated by a modest number of other factors, the problem is exponential in the number of inputs per gene, but only polynomial in the number of genes. So we have a real chance at cracking the inverse problem. I think it's going to be one of the most important things that will happen in the next 15 years.

SA: Is the inverse problem a true problem or one method to get information?
It's a true problem. The direct problem is that I write down the equations for some dynamical system and it then goes and behaves in its dynamical way. The inverse problem is you look at the dynamical behavior and you try and figure out the laws. The inverse problem for us is we see the dynamical behavior of the genome displaying itself on Affymetrix chips or proteomic display methods, like 2–D gels. And now we want to deduce from that what the circuitry and the logic is. So it's the general form of a problem. It's the way to try and get out which genes are regulating which genes, so that I know not just from my ensemble approach, but I know that gene A really is regulating gene F.

SA: *What are the barriers to figuring out the inverse problem? Is it computer power? Designing the proper algorithms? Biomolecular understanding?*

All three. Let's take a case in point: Feedback loops make it hard to figure things out. And the genome is almost certainly full of feedback loops. For example, there are plenty of genes that regulate their own activity. So figuring out the algorithms that will deal with feedback loops is not going to be trivial. The computing power gets to be large because if I want to look for a single input, like a canalizing input, the canalizing input is really easy to tell because if gene A is on, then gene C is on no matter what. So all I have to do is examine a lot of gene expression data and I can see whenever A is on, now C is on a moment later.

I can do that by looking at things one gene at a time. But suppose I had a more complicated rule in which two genes had to be in a combined state activity to turn on gene C. To do that, I have to look at genes pair-wise to see that they manage to regulate gene C. If I looked at A alone or B alone, I wouldn't learn anything. So now if I have 100,000 genes, I've got to look at $100,000^2$ pairs; that's 10^{10} pairs. Now what if I have a rule that depends on the state of activity of three genes to turn the gene on, then I have to look at $100,000^3$, which is 10^{15}, and that's probably about the limit of the computing power that we've got now.

But that leaves out the fact that we don't have to be stubborn about it; we can always go do an experiment. And so this now ties to experimental design. Notice that all of these problems lead in the direction of new experimental designs, and what we're going to have to do is to marry things like the inverse problem to being able to toggle the activity of arbitrary genes.

SA: *What about the molecular biology knowledge?*

It's going to take a lot of biological knowledge. For example, let's suppose that every structural gene has at least one cis site that regulates it. Then there's 100,000 cis sites. Nobody knows if that's true, but let's pose that. Well, we have an awful lot of work to do to pull out all the cis sites. Now let's suppose that 5 percent of the structural genes that are around act as regulatory inputs to the structural genes. So there's on the order of 4,000 to 5,000 trans factors making up this vast network that we're talking about.

Well, we have to discover what those trans factors are; we have to discover what the cis sites are; we have to discover what the regulatory logic is. Then we have to make mathematical models of it. Then we have to integrate the behavior of those mathematical models. And then we're going to run into the same problems that people have looking at the gut ganglion in lobster—that even though you know all the inputs, figuring out the behavior is going to be hard. Then we're going to run into the problem that you're looking at a circuitry with 40 genes in it, but there are impacts coming in from other genes in the 100,000–gene network that are going to screw up your models. So, this ain't going to be easy.

SA: *In terms of a timeline, are we looking at 1 year or 200 years away?*
I think 30 to 40 years from now we will have solved major chunks of this. The tools will mature in the next 10 to 12 years, and then we'll really start making progress.

SA: *What do you define as progress?*
That we will be getting the circuitry for big chunks of the genome and actually understanding how it works. Getting to the genomic sequences is wonderful, but what does it tell you about the circuitry? So far, nothing—except who the players are.

SA: *So we're at the beginning?*
We're at the very beginning of postgenomic medicine. But the payoff is going to be enormous. There's going to be a day 30

years from now where somebody comes in with cancer and we diagnosis it with accuracy not just on the morphology of the cancer cell but by looking at the detailed patterns of gene expression and cis site binding activities in that cell. And we know the circuitry and the autocrine perturbations to try, or we know which gene activity perturbations to try that will cause that cell to differentiate into a normal cell type or cause that cell to commit hara-kiri.

SA: Will that be someone walking into their doctor's office, the doctor turning on the computer and just entering the data?

It will require being able to do the RNA sample. Biotech companies together with big pharma, which alone has the money to get things through clinical trials, will wind up proving that you can treat cancer this way—or treating some degenerative disease of your joint, for example, where we regenerate the synovium. Why not? We can make an entire sheep—why can't we regenerate the synovium?

The patenting of genes is such a recent phenomenon that the protocol has not be established. Some companies are not only attempting to get a patent for the sequence of a given gene, they are also seeking a patent for the computer codes used to store the data of such sequences. But if a company were allowed such a blanket patent, would it hamper scientific progress?

Code of the Code

Gary Stix

I n 1995 Craig Venter and his colleagues at the Institute for Genomic Research (TIGR) became the first to sequence all the A, G, C and T nucleotides in the genome of a free-living organism—the bacterium *Hemophilus influenzae*, which causes ear and respiratory infections. Human Genome Sciences (HGS), a major biotechnology firm with which TIGR was affiliated at the time, applied for a patent not just on the sequence of nucleotides in the DNA itself but on any "computer-readable medium having recorded thereon the nucleotide sequence." In essence, the application asked for a patent on the exclusive use of the computer code representing the germ's genetic code.

The patent, which is still pending in the U.S. and elsewhere, represents a "fundamental departure" from previous practice, wrote biotechnology law scholar Rebecca Eisenberg in 2000 in the *Emory Law Journal*: "By claiming exclusionary rights in the sequence information itself, if stored in a computer-readable medium, HGS seeks patent rights that would be infringed by information storage, retrieval and analysis rather than simply by making, using or selling DNA molecules." HGS and at least one other company have filed similar applications on other genomes, but it is highly uncertain that the U.S. Patent and Trademark Office will approve them, as it has repeatedly tightened rules to prevent patenting of genes for which there are no clearcut uses.

Even if these patents are denied, though, the blurring of distinctions between molecular and digital information is very likely to continue. Companies might seek protection for the code of a three-dimensional computerized representation of a receptor on a cell. And patents related to information gleaned from gene chips—which use segments of DNA as detectors to determine the presence of genes expressed in a given sample— pose similar dilemmas.

Such patents would have potentially far-reaching consequences. If accessing a patent on the Internet were to constitute an infringement, this would go against the fundamental quid pro quo on which patent law is based, Eisenberg contends. The holder of a patent gets a 20-year monopoly on the right to make, use and sell an invention in exchange for revealing information about both its manufacture and usage. Access to this information promotes the free exchange of ideas essential to technological progress. "If the terms of the traditional patent bargain are altered to allow patent holders to capture the informational value of their discoveries," Eisenberg writes, "the bargain becomes less attractive to the public." Others cannot avail themselves of information needed to enhance the state of the art. If DNA as information exceeds its value as a tangible molecule, it may be necessary to find some other intellectual-property protection for it. Patenting the zeros and ones representing As, Gs, Cs and Ts won't cut it.

Sequencing the human genome has opened the floodgates of research. What's next? Researchers are now paying attention to the body of messenger RNA being produced by a cell at any given time and all of the proteins being made according to the instruction in those messenger RNAs. One new industry to come from this initial research is proteomics, the study of protein expression, which would help drug designers to create drugs that either activate proteins or prevent them from interacting.

Beyond the Human Genome

Carol Ezzell

Genes are all the rage right now, but in a sense, at this very moment, they are also becoming passé. Now that all the genes that make up the human genome have been deciphered, a new industry is emerging to capitalize on when and where those genes are active and on identifying and determining the properties of the proteins the genes encode. The enterprise, which has so far attracted hundreds of millions of dollars in venture capital and other financing, can be lumped under the newly coined term "proteomics."

"The biggest issue for genomics today is no longer genes," asserts William A. Haseltine, chairman and chief executive officer of Human Genome Sciences in Rockville, Md. "What's interesting is what you do with those genes."

"We have to move on to understand the other elements of the biological process and couple all this [information] together," agrees Peter Barrett, chief business officer of Celera Genomics, also in Rockville, the company that raced the publicly funded Human Genome Project to sequence the human genome. "People took it for granted that the [human] genome would be done in 2000. Now it's 'What do we do next?'"

What's next, for the most part, are messenger RNAs (mRNAs) and proteins. If DNA is the set of master blueprints a cell uses to construct proteins, then mRNA is like the copy of part of the blueprint that a contractor takes to the building site every day. DNA remains in the nucleus of a cell; mRNAs transcribed from active genes leave the nucleus to give the orders for making proteins.

Although every cell in the body contains all of the DNA codes for making and maintaining a human being, many of those genes are never "turned on," or copied into mRNA, once embryonic development is complete. Various other genes are turned on or off at different times—or not at all—according to the tissue they are in and their role in the body. A pancreatic beta cell, for instance, is generally full of the mRNA instructions for making insulin, whereas a nerve cell in the brain usually isn't.

Scientists used to think that one gene equals one mRNA equals one protein, but the reality is much more complicated. They now know that one gene can be read out in portions that are spliced and diced to generate a variety of mRNAs and that subsequent processing of the newly made proteins that those transcripts encode can alter their function. The DNA sequence of the human genome therefore tells only a small fraction of the story about what a specific cell is doing. Instead, researchers must also pay attention to the transcriptome—the body of mRNAs being produced by a cell at any given time—and the proteome, all the proteins being made according to the instructions in those mRNAs.

Cashing in on Chips

One of the technologies for studying the human transcriptome is the GeneChip system developed by Affymetrix in Santa

Clara, Calif. The system is based on thumbnail-size glass chips called microarrays that are coated with a thin layer of so-called cDNAs, which represent all the mRNAs made by a particular type of cell. (The abbreviation cDNA stands for "complementary DNA;" it is essentially mRNA artificially translated back into DNA, but without the noncoding sequence gaps, or introns, found in the original genomic DNA.)

To use the system, scientists isolate mRNA from their cellular sample, tag it with a chemical marker and pour it over the chip. By observing where the sample mRNA matches and binds to the cDNA on the chip, they can identify the mRNA sequences in their sample. In 2000 Affymetrix launched two new sets of chips for analyzing human cell samples. One allows researchers to identify more than 60,000 different human mRNAs; the other can screen cells for roughly 1,700 human mRNAs related to cancer.

The National Cancer Institute in Bethesda, Md., has been examining the mRNAs produced by various types of cancer cells for years now, in a project called the Human Tumor Gene Index. The index is a partnership between government and academic laboratories as well as a group of drug companies that includes Bristol-Myers Squibb, Genentech, Glaxo Wellcome and Merck. So far they have identified more than 50,000 genes that are active in one or more cancers. For instance, the index has found that 5,692 genes are active in breast cancer cells, including 277 that are not active in other tissues. Compounds that home in on the proteins produced by those 277 genes might serve as good cancer drugs with fewer side effects than current chemotherapies. The National Cancer Institute has also recently begun a multimillion-dollar Tissue Proteomics Initiative in conjunction with the U.S. Food and Drug Administration to identify proteins involved in cancer.

At bottom, mRNA studies are just a means to better understand the proteins in a cell's production line—after all, the proteins are the drug targets. And with researchers expecting that the human genes will turn out to produce more than a million proteins, that's a lot of targets. Jean-François Formela of Atlas Venture in Boston estimates that within the next decade the pharmaceutical industry will be faced with evaluating up to 10,000 human proteins against which new therapeutics might be directed. That's 25 times the number of drug targets that have been evaluated by all pharmaceutical companies since the dawn of the industry, he says.

Mark J. Levin, CEO of Millennium Pharmaceuticals in Cambridge, Mass., says that large pharmaceutical companies, or "big pharma," need to identify between three and five new drug candidates a year in order to grow 10 to 20 percent a year—the minimum increase shareholders will tolerate. "Right now the major pharma companies are only delivering a half to one-and-a-half entities a year," Levin explains. "Their productivity will not sustain their ability to continue to develop and create shareholder value." Millennium has a relationship with Bayer to deliver 225 pretested "druggable" targets within a few years.

"Protein expression is now capturing the imagination of scientists," comments Randall W. Scott, chief scientific officer of Incyte Genomics in Palo Alto, Calif. "It's being able to look not just at a gene and how it's expressed, but at the forms of the protein."

Protein Machines

Scientists at the DNA-sequencing juggernaut Celera are among those getting interested in the study of protein expression, or proteomics. Celera has been in negotiations with GeneBio, a commercial adjunct of the Swiss Institute for Bioinformatics in Geneva, to launch a company dedicated to deducing the entire human proteome. In 1999 Denis F. Hochstrasser, one of the founders of GeneBio, and his colleagues published plans for a molecular scanner that would automate the now tedious process of separating and identifying the thousands of protein types in a cell.

The current method for studying proteins consists in part of a technique called two-dimensional gel electrophoresis, which separates proteins by charge and size. In the technique, researchers squirt a solution of cell contents onto a narrow polymer strip that has a gradient of acidity. When the strip is exposed to an electric current, each protein in the mixture settles into a layer according to its charge. Next, the strip is placed along the edge of a flat gel and exposed to electricity again. As the proteins migrate through the gel, they separate according to their molecular weight. What results is a smudgy pattern of dots, each of which contains a different protein.

In academic laboratories, scientists generally use a tool similar to a hole puncher to cut the protein spots from 2-D gels for individual identification by another method, mass spectroscopy. But the companies Large Scale Biology in Vacaville, Calif., and Oxford GlycoSciences (OGS) in Oxford, England, use robots to do it. OGS is under contract with Pfizer to analyze samples of cerebrospinal fluid taken from patients with various stages of Alzheimer's disease.

The machine devised by Hochstrasser and his research group goes one step further than the robots used by Large Scale Biology and OGS. It would automatically extract the protein spots from the gels, use enzymes to chop the proteins into bits, feed the pieces into a laser mass spectrometer and transfer the information to a computer for analysis. The instrument manufacturer PE Corporation, which owns Celera, has already agreed to make the machines.

With or without robotic arms, 2-D gels have their problems. Besides being tricky to make, they don't resolve highly charged or low-mass proteins very well. They also do a poor job of resolving proteins with hydrophobic regions, such as those that span the cell membrane. This is a major limitation, because membrane-spanning receptors are important drug targets.

Another method for studying proteomes is what Stephen Oliver of the University of Manchester in England has called "guilt by association": learning about the function of a protein by assessing whether it interacts with another protein whose role in a cell is known. In February 2000 researchers at Cura-Gen in New Haven, Conn.—together with a group led by Stanley Fields of the Howard Hughes Medical Institute at the University of Washington—reported that they had deduced 957 interactions among 1,004 proteins in the baker's yeast *Saccharomyces cerevisiae*. Fields and his colleagues first devised a widely used method for studying protein interactions called the "yeast two-hybrid system", which uses known protein "baits" to find "prey" proteins that bind to the "baits."

The yeast genome has been known to consist of 6,000 genes since it was sequenced in 1996, but the functions of one-third of them have remained mysterious. By figuring out which of the unknown proteins associated with previously identified ones, the CuraGen and University of Washington scientists were able to sort them into functional categories, such as energy generation, DNA repair and aging.

In March 2000 CuraGen announced that it had teamed up with the Berkeley Drosophila Genome Project to produce a protein-interaction map of the fruit fly. "We want to take this massively parallel approach forward," says Jonathan M. Rothberg, CuraGen's founder, chairman and CEO. The director of the Berkeley project is Gerald M. Rubin, a Howard Hughes Medical Institute researcher at the University of California at Berkeley. He collaborated with Celera on the sequencing of the *Drosophila* genome. "Yeast was a prototype for us," Rothberg explains. "But *Drosophila* is a good choice when you want to study an organism with multiple cells." CuraGen aims to use proteomics to find new drugs for its clients to bring to market. "Our proteomics is 100 percent 'What does your gene do?' and 'Is it a drug target?'" Rothberg states. But CuraGen will also work to identify targets for drugs to sell on its own.

One of CuraGen's competitors is Myriad Genetics, a biotechnology company based in Salt Lake City that is best known for its tests for the *BRCA* genes that contribute to breast and ovarian cancer. Myriad made a deal worth up to $13 million with Roche to lend its proteomics techniques to finding targets for potential cardiovascular disease drugs.

Myriad also uses a variation of the yeast two-hybrid system but concentrates on specific disease pathways rather than assessing entire genomes. The company has an ongoing alliance with Schering-Plough, for instance, to plumb the biochemical interactions of proteins encoded by a gene called *MMAC1*, which when mutated can lead to brain and prostate cancer.

Another way to study proteins that has recently become available involves so-called protein chips. Ciphergen Biosystems, a biotechnology company in Palo Alto, California, is now selling a range of strips for isolating proteins according to various properties, such as whether they dissolve in water or bind to charged metal atoms. The strips can then be placed in Ciphergen's chip reader, which includes a mass spectrometer, for identifying the proteins.

One of the initial uses of Ciphergen's protein chips has been in finding early markers for prostate cancer. George L. Wright, Jr., of Eastern Virginia Medical School in Norfolk reported using Ciphergen's system to identify 12 candidate "biomarkers" for benign prostatic disease and six such biomarkers for prostate cancer. Tests based on the proteins might be better at discriminating between benign and cancerous prostate conditions than the currently available prostate specific antigen (PSA) assay.

The Structure's the Thing

Identifying all of the proteins in a human is one thing, but to truly understand a protein's function scientists must discern its shape and structure. In an article in *Nature Genetics* in October 1999, a group of well-known structural biologists led by Stephen K. Burley of the Rockefeller University called for a "structural genomics initiative" to use quasi-automated x-ray crystallography to study normal and abnormal proteins.

Conventional structural biology is based on purifying a molecule, coaxing it to grow into crystals and then bombarding the sample with x-rays. The x-rays bounce off the molecule's atoms, leaving a diffraction pattern that can be interpreted to yield the molecule's overall three-dimensional shape. A structural genomics initiative would depend on scaling up and speeding up the current techniques.

The National Institutes of Health is poised to award $20 million in grants for structural genomics to academic centers. And companies are getting into the game, too: Syrrx in La Jolla, Calif., Structural GenomiX in San Diego, California, and Chalon Biotech in Toronto are founded on developing so-called high-throughput x-ray crystallographic techniques.

Knowing the exact structural form of each of the proteins in the human proteome should, in theory, help drug designers devise chemicals to fit the slots on the proteins that either activate them or prevent them from interacting. Such efforts, which are generally known as rational drug design, have not shown widespread success so far—but then only roughly one percent of all human proteins have had their structures determined. After scientists catalogue the human proteome, it will be the proteins—not the genes—that will be all the rage.

As the final essays tell us, humans are not the only organisms on earth with genomes! The vast majority of earth's life forms employ DNA to transport their genetic information from generation to generation, so each species is potentially the focus of a Genome Project all its own. At the moment, however, sequencing the genome of any other complex organism would be as expensive and as time-consuming as the Human Genome Project. So for a while, the organisms with completely sequenced genomes will be those with simple genomes and those with scientific or economic importance, such as laboratory animals and agricultural animals and plants. But eventually we can expect a breakthrough in the speed with which any genome can be sequenced. In time, it might even become possible to save species from extinction by preserving their genomes in a huge life-forms database from which they could be resurrected at will.

One of the surprising recent results of the Human Genome Project is that we have less than half the number of genes we thought we had: merely four times as many genes as the celebrated fruit fly. Does this mean that humans are just four times as complex as fruit flies? Probably not: complexity has a way of increasing much faster than the sum or product of the number of components of a system. Rather, it means that being multicellular itself requires a certain level of complexity, and that being a multicellular something is merely a kind of embellishment on this fundamental multicellularity. It also suggests that quite a bit of the genome is devoted to regulating an organism's development from egg to adult, so that what makes us different from insects and worms lies in a large number of comparatively minor changes to this regulatory process. We may thus be closer to our fellow life forms than we suspect.

The "Other" Genomes

Julia Karow

W hat do we have in common with flies, worms, yeast and mice? Not much, it seems at first sight. Yet corporate and academic researchers are using the genomes of these so-called model organisms to study a variety of human diseases, including cancer and diabetes.

The genes of model organisms are so attractive to drug hunters because in many cases the proteins they encode closely resemble those of humans—and model organisms are much easier to keep in the laboratory. "Somewhere between 50 and 80 percent of the time, a random human gene will have a sufficiently similar counterpart in nematode worms or fruit flies, such that you can study the function of that gene," explains Carl D. Johnson, vice president of research at Axys Pharmaceuticals in South San Francisco, California.

Here's a rundown on the status of the genome projects of the major model organisms today:

The Fruit Fly

The genome sequence for the fruit fly *Drosophila melanogaster* was completed in March 2000 by a collaborative of academic investigators and scientists at Celera Genomics in Rockville, Md.

The researchers found that 60 percent of the 289 known human disease genes have equivalents in flies and that about 7,000 (50 percent) of all fly proteins show similarities to known mammalian proteins.

One of the fly genes with a human counterpart is *p53*, a so-called tumor suppressor gene that when mutated allows cells to become cancerous. The *p53* gene is part of a molecular pathway that causes cells that have suffered irreparable genetic damage to commit suicide. In March 2000 a group of scientists, including those at Exelixis in South San Francisco,California, identified the fly version of *p53* and found that—just as in human cells—fly cells in which the *p53* protein is rendered inactive lose the ability to self-destruct after they sustain genetic damage and instead grow uncontrollably. Similarities such as this make flies "a good trade-off" for studying the molecular events that underlie human cancer, according to one of the leaders of the fly genome project, Gerald M. Rubin of the Howard Hughes Medical Institute at the University of California at Berkeley: "You can do very sophisticated genetic manipulations [in flies] that you cannot do in mice because they are too expensive and too big."

The Worm

When researchers deciphered the full genome sequence of the nematode *Caenorhabditis elegans* in 1998, they found that roughly one third of the worm's proteins—more than 6,000— are similar to those of mammals. Now several companies are taking advantage of the tiny size of nematodes—roughly one millimeter—by using them in automated screening tests to search for new drugs.

To conduct the tests, scientists place between one and ten of the microscopic worms into the pill-size wells of a plastic microtiter plate the size of a dollar bill. In a version of the test used to screen for diabetes drugs, the researchers use worms that have a mutation in the gene for the insulin receptor that causes them to arrest their growth. By adding various chemicals to the wells, the scientists can determine which ones restore the growth of the worms, an indication that the compounds are bypassing the faulty receptor. Because the cells of many diabetics no longer respond to insulin, such compounds might serve as the basis for new diabetes treatments.

The Yeast

The humble baker's yeast *Saccharomyces cerevisiae* was the first organism with a nucleus to have its genetic secrets read, in 1996. Approximately 2,300 (38 percent) of all yeast proteins are similar to all known mammalian proteins, which makes yeast a particularly good model organism for studying cancer: scientists first discovered the fundamental mechanisms cells use to control how and when they divide using the tiny fungus.

"We have come to understand a lot about cell division and DNA repair—processes that are important in cancer—from simple systems like yeast," explains Leland H. Hartwell, president and director of the Fred Hutchinson Cancer Research Center in Seattle and co-founder of the Seattle Project, a collaboration between academia and industry. So far Seattle Project scientists have used yeast to elucidate how some of the existing cancer drugs exert their function. One of their findings is that the common chemotherapeutic drug cisplatin is particularly effective in killing cancer cells that have a specific defect in their ability to repair their DNA.

The Mouse

As valuable as the other model organisms are, all new drugs must ultimately be tested in mammals—and that often means mice. Mice are very close to humans in terms of their genome: more than 90 percent of the mouse proteins identified so far show similarities to known human proteins. Ten laboratories across the U.S., called the Mouse Genome Sequencing Network, collectively received $21 million from the National Institutes of Health in 1999 to lead an effort to sequence the mouse genome. At this writing they have completed approximately nine percent of it, and their goal is to have a rough draft ready by 2003. But that timeline might be sped up: Celera announced in April 2000 that it is turning its considerable sequencing power to the task.

Reading the
Book of Life

Julia Karow

I n the summer of 2000, the world celebrated when scientists from the Human Genome Project, an international consortium of academic research centers, and Celera Genomics, a private U.S. company, both announced that they had finished working drafts of the human genome. It was an important first step toward deciphering the entire genome, one of the greatest scientific undertakings of all time. But these drafts revealed only the beginning of the story—the scrolls containing the instructions for life. Now both teams have started reading—gene after gene—the actual scriptures within the scrolls.

Among other surprises, both teams agree that humans have a mere 26,000 to 40,000 genes—which is far fewer than many people predicted. For perspective, consider that the simple roundworm *Caenorhabditis elegans* has 18,000 genes; the fruit fly *Drosophila melanogaster,* 13,000. Some estimated the human genome might include as many as 140,000 genes. It will be several more years before scientists agree on an absolute total, but most are confident that the final number won't fall out of the range reported. "I wouldn't be shocked if it was 29,000 or 36,000," says Francis Collins, director of the National Human Research Institute at the NIH. "But I would be shocked if it was 50,000 or 20,000."

An error margin of some 10,000 genes may not seem impressive after so many years of work, but genes—the actual units of DNA that encode RNA and proteins—are very difficult to count. For one thing, they are scattered throughout the genome like proverbial needles in a haystack: their coding parts constitute only about 1 to 1.5 percent of the roughly three billion base pairs in the human genome. The coding region of a gene is fragmented into little pieces, called exons, linked by long stretches of non-coding DNA, or introns. Only when messenger RNA is made during a process called transcription are the exons spliced together.

To identify functional genes, Collins explains, the scientists had to "depend upon a variety of bits of clues." Some clues come from comparisons with databases of complementary DNAs (cDNAs), which are exact copies of messenger RNAs. So, too, comparisons with the mouse genome help because most mouse and human genes are very similar; their sequences are conserved in both genomes, whereas a lot of the surrounding DNA is not. And when such clues aren't available, scientists rely exclusively on gene-predicting computer algorithms.

Because these algorithms are not totally reliable—sometimes they see a gene where there is none or miss one altogether—a few scientists doubt the new human gene count. For instance, William Haseltine of Human Genome Sciences—a company that specializes in finding protein-encoding genes only on the basis of cDNA—thinks that "the methods that have been used are very crude and inexact." He believes that there are more than twice as many genes as reported thus far by the two groups.

But many others do accept the current estimates and are asking what it means that humans should have so few genes.

According to Craig Venter, president of Celera Genomics, "the small number of genes means that there is not a gene for each human trait, that these come at the protein level and at the complex cellular level." As it turns out, at least every third human gene makes several different proteins through "alternative splicing" of its pre-messenger-RNA. Also human proteins have a more complicated architecture than their worm and fly counterparts, adding another level of complexity. And compared with simpler organisms, humans possess extra proteins having functions, for example, in the immune system and the nervous system, and for blood clotting, cell signaling and development.

Scientists are also puzzling over the significance of the discovery that more than 200 genes from bacteria apparently invaded the human genome millions of years ago, becoming permanent additions. The new work shows that some of these bacterial genes have taken over important human functions, such as regulating responses to stress. "This is kind of a shocker and will no doubt inspire some further study," Collins says. Indeed, scientists previously thought that this kind of horizontal gene transfer was not possible in vertebrates.

Another curious feature of the human genome is its overall landscape, in which gene-dense and gene-poor regions alternate. "There are these areas that look like urban areas with skyscrapers of gene sequences packed on top of each other," Collins explains, "and then there are these big deserts where there doesn't seem to be anything going on for millions of base pairs." Moreover, such differences are apparent not only within ,but also between chromosomes. Chromosome 19, for example, is about four times richer in genes than the Y-chromosome.

So what's going on in gene deserts? More than half the human genome consists of repeat sequences, also known as "junk DNA" because they have no known function. Vertebrates can live well without them: the puffer fish, for example, has a genome with very few of these repeats. In humans, most of them derive from transposable elements, parasitic stretches of DNA that replicate and insert a copy of themselves at another site. But now almost all the different families of transposons seem to have stopped roaming the genome, and only their "fossils" remain. Still, nearly 50 genes appear to originate from transposons, suggesting they played some useful role during the genome's evolution.

One type of transposon, the so-called Alu element, is found especially often in regions rich in G and C bases. These areas also harbor many genes, and so Alu's might somehow be beneficial around them. Overall, the human genome once seemed to be "a complex ecosystem, with all these different elements trying to proliferate," says Robert Waterston, director of the Genome Sequencing Center at the University of Washington, a member of the public consortium. The mutations they have accumulated provide an excellent molecular fossil record of the evolutionary history of humankind.

In addition to repeat sequences caused by transposons, large segments of the genome seem to have duplicated over time, both within and between chromosomes. This duplication, researchers say, allowed evolution to play with different genes without destroying their original function and probably led to the expansion of many gene families in humans.

Apart from the genome sequence, both the Human Genome Project and Celera have identified a multitude of base positions in the DNA that differ between individuals and are called single polynucleotide polymorphisms, or SNPs (pronounced "snips"). The public consortium discovered 1.4 million SNPs, and Celera announced it had found 2.1 million of them. Scientists are hoping to learn from them how genes make people different and, in particular, why some are more susceptible to certain diseases than others. "It will certainly take us a long time to figure out what they all mean, if they all mean anything, but I think the process is already beginning," Waterston notes.

To be sure, much work remains. Only one billion base pairs, a third of the total, in the public database are in a "finished" form, meaning they are highly accurate and without gaps. Both the Celera and the public data contain numerous gaps. In addition, large parts of the heterochromatin—a gene-poor, repeat-rich part of the DNA that accounts for about ten percent of the genome—has yet to be cloned and sequenced. By the spring of 2003, the public project is hoping to finish that task, except for sequences that turn out to be impossible to obtain using current methods.

The next big challenge will be to find out how the genes interact in a cell. According to Collins, researchers will "begin to look at biology in a whole-genome way," studying, for example, the expression of all genes in a cell at a given time. Proteins, the products of the genes, will also be studied "not just one at a time, but tens of thousands at a time," Collins says, speaking of a fast-growing research field that goes by the name of proteomics. In the end, however, genes may provide only so many answers. "The Basic message," Venter concludes, "is that humans are not hardwired. People who were looking for deterministic explanations for everything in their lives will be very disappointed, and people who are looking for the genome to absolve them of personal responsibility will be even more disappointed."

all new drugs must ultimat
ted in mammals—and that ofte
e. Mice are very close to h
ms of their genome: more tha
cent of the mouse proteins i
far sho Conclusion es to kno
teins. Ten laboratories acr
called the Mouse Genome Sec
work, collectively received
m the National Institiutes o

Conclusion

While the draft of the human genome is a culmination of more than a decade of research, it will also serve as the springboard for a myriad of overlapping discoveries and biomedical innovations for many decades to come.

The results of these efforts will touch every facet of our lives. Scientists will build on this new-found knowledge to treat diseases by manipulating genes and their functions and ultimately to prevent them. Anthropologists will use it to fill in gaps in the story of our species evolution.

But a number of hard questions have also been created by the decoding. One big concern is the legal ownership of the human genome. The nascent race to patent specific genes has raised the conflicting issues of the proprietary rights of discovery versus the need for open access to information that is essential to building a robust biotech industry. Without patents, the private sector will be reluctant to pay the funding needed to stimulate the rapid development of genome-based health care products. With patents, a few organizations may end up controlling an invaluable resource. At the rate of progress in gene research, this conflict is destined to come to a head soon.

Individuals will also face a dilemma between their right to learn about a predisposition to an illness versus the fear of being discriminated against for this predisposition by insurance companies and employers. People who learn that they have genes that allow for a disease will also have to tackle the psychological ramifications of knowing their potential future.

But even with these challenges, the opportunity for progress is too great to ignore. Each discovery and innovation inevitably leads to another. For example, recent studies have shown that gene activity, not sequence, could be the key to our identity. The differences in gene expression may explain how largely similar genomes create such different organisms. Therefore the future of gene research may lie in the Human Proteome Project, which has been established to identify the estimated 300,000 to 2 million proteins in the human body. This further unraveling of the genomic mystery will ultimately allow the biomedical community to create a new evolutionary future for the human race. And biomedicine will never be the same. As for humanity. . .well, we'll just have to wait and see. But the wait doesn't feel like it will be long at all.

Index

Photo Credits

Page 28, Everard Williams, Jr.. Page 30-31, Bryan Christie. Page 34 Bryan Christie, with assistance from John Logsdon, Dalhousie University. Page 42-43 Jennifer C. Christiansen. Page 46 Jennifer C. Christiansen. Page 92 Laurie Grace, with assistance from Mark Gerstein and Pat Fleming, Yale University and David Wheeler and Jennifer Vyskocil, NCBI.